ENGINEERING
EXCELLENCE
CULTURAL AND ORGANIZATIONAL FACTORS

OTHER IEEE PRESS BOOKS

Writing Reports to Get Results, *by R. S. Blicq*
Effective Meetings for Busy People, *by W. T. Carnes*
Being the Boss, *by L. K. Lineback*
Getting the Picture, *by S. B. Weinstein*
Space Science and Applications, *Edited by J. H. McElroy*
The Calculus Tutoring Book, *by C. Ash and R. Ash*
Next-Generation Computers, *Edited by E. A. Torrero*
Insights Into Personal Computers, *Edited by A. Gupta and H. D. Toong*
The Space Station: An Idea Whose Time Has Come, *by T. R. Simpson*
The Making of a Profession: A Century of Electrical Engineering in America, *by A. M. McMahon*
General and Industrial Management, *by H. Fayol; revised by I. Gray*
Engineers and Electrons, *by J. D. Ryder and D. G. Fink*
Compendium of Communication and Broadcast Satellites: 1958 to 1980, *Edited by M. P. Brown, Jr.*
The Engineer in Transition to Management, *by I. Gray*

ENGINEERING EXCELLENCE

CULTURAL AND ORGANIZATIONAL FACTORS

Donald Christiansen
Editor

IEEE PRESS

The Institute of Electrical and Electronics Engineers, Inc., New York

PRINTED IN THE UNITED STATES OF AMERICA

IEEE Order Number: PC02188

Library of Congress Cataloging-in-Publication Data

Engineering excellence—cultural and organizational factors.

Includes index.
1. Engineering—United States. 2. Engineering—Europe. 3. Engineering—Japan. I. Christiansen, Donald.
TA157.E613 1987 306′.46 87-3626
ISBN 0-87942-229-7

CONTENTS

PREFACE

The country that trains its engineers and technologists well, then rewards them with both real and psychic income, should have little trouble competing in a world economy that thrives on trading high quality, high tech products over international boundaries. Or so it might seem. However, this worthy goal is easier to state than to achieve. Furthermore, there are political, economic, and organizational factors that affect and sometimes override efforts to exploit outstanding technical developments.

In the environment of recent years, and particularly that of today, despite ostensible good will on the part of world trading partners, the desired goals are elusive. Among the concerns are trade imbalances, the loss of markets and skills on the part of countries who find themselves unable to compete in selected markets, and poor quality products.

In the United States, emotions have run high as Japanese industry was believed to be selling semiconductor chips below cost in order to gain a greater share of market compared to their U.S. and European competitors. The U.S. government weighed the impact of taking responsive actions and adjusting them as it deemed appropriate.

It was in this general environment that the Institute of Electrical and Electronics Engineers, under the leadership of its 1986 president, Bruno O. Weinschel, encouraged prominent industry, government, and academic leaders to come together in San Jose, California, on February 18–19, 1986, to consider how cultural traditions in their respective countries might relate to ''engineering excellence'' as a desired characteristic, and to the way in which engineers are managed and rewarded. The participants in the resulting convocation represented Japan, Germany, Holland, and the United States, and the substance of their presentations forms the major part of this volume.

Not by design, but nevertheless somewhat naturally, the presentations fall into three categories: The importance of the individual engineer, a comparison of engineering cultures, and institutional and organizational factors. As editor of this volume, I invited a post-convocation discussion of each group of articles.

Part I is a grouping of papers covering the importance of the individual engineer, including chapters by Masaru Ibuka, Honorary Chairman and Founder of the Sony Corporation, Piet Kramer, Senior Managing Director of Corporate Research at Philips International B.V., Dean O. Morton, Executive Vice President and Chief Executive Officer of the Hewlett-Packard Company, and Robert N. Noyce, Vice Chairman of the Board of Intel Corporation. Michael F. Wolff, Editor of *Research Management* and a scholar on the process of innovation, discusses these papers in a summary chapter.

Part II includes several papers that compare engineering cultures. Authors include Karl H. Beckurts, former Executive Vice President and Head of Corporate Technology at Siemens AG, Myron Tribus, former director of the Center for Advanced Study at the Massachusetts Institute of Technology, Yoshi Tsurumi, Professor of International Business at Baruch College in the City University of New York, and Michiyuki Uenohara, Executive Vice President and Director of the NEC Corporation. George Wise, a professional historian of technology and Corporate Historian for the General Electric Company, discusses this set of papers.

Part III treats institutional and organizational factors, with papers by Frank L. Huband, Director of the Electrical, Communications, and Systems Engineering Division of the National Science Foundation, and James D. Meindl, Vice President of Academic Affairs and Provost of the Rensselaer Polytechnic Institute. Gene W. Dalton and Lee Tom Perry, professors of organizational behavior at Brigham Young University's Graduate School of Management, discuss the two papers in this section in a summary chapter.

Part IV includes several papers that look at the issues of engineering excellence and managing technology from other viewpoints. It includes chapters by, among others, Roland W. Schmitt, Senior Vice President and Chief Scientist at General Electric Company, Professors Robert H. Hayes and William J. Abernathy of the Harvard Business School, Prof. Hans Kornberg of the University of Cambridge and Lawrence P. Grayson, an advisor to the U.S. Department of Education.

Part of the hoped-for result of the convocation was a recognition of those goals and methods that represent a common thread among industrialized countries. We were also hoping to identify certain successful organizational and management techniques that are culturally driven but not necessarily limited in application to their country of origin. The reader, I believe, will find a number of such transferable ideas.

It is possible, too, that this volume may help authenticate or disprove some of the myths concerning international competition in our high tech world, or at least encourage further discussion in that direction. There has been, for example, a general belief in the United States that trade unions play no significant role in Japanese industry. Another is that Japanese workers have a job for life. Still another is that Japanese engineers (and other "white collar" workers in Japan) work diligently and happily until 7 or 8 p.m. daily. None of these is strictly true, and with the passage of time problems related to some of these areas that have already plagued U.S. industry are becoming evident in Japan, too. Nevertheless, it is enlightening to learn of the current practices and motivations of engineers in other industrialized countries.

Monetary and psychic rewards for engineers vary significantly from one culture to another. Interestingly, engineers outside the United States traditionally list their degrees and professional societies on their business cards while many U.S. engineers consider that practice unusual if not ostentatious. On the other hand, awards to individuals for outstanding professional contributions, generally highly esteemed in the United States, are not universally so considered. In some countries they are deemed ostentatious, or at least to be accepted humbly and without fanfare.

Likewise, whereas the recognition of accomplishments by young engineers is common and even institutionalized in the United States (e.g., Eta Kappa Nu's Outstanding Young Electrical Engineer Award), engineers in other countries are expected to "pay their dues" in terms of years of service before being eligible for promotions, honors, and offices.

It is virtually impossible to discuss engineering excellence without also discussing the methods of managing and motivating engineers. Thus, most of the writers in this volume deal with management either directly or indirectly.

Corporate managers are both driven and constrained by the economic system in which they manage. U.S. managers must react in the short term to stockholder pressures. Worse, they may find it necessary to take drastic actions not in the long-term best interests of the organization in order to fend off corporate takeovers. Such is not the case elsewhere, notably in Japan.

Except in the United States, where opinion polls are accepted as the norm, statistical data on how engineers perceive their job environments or their managers are scarce.

A recent study by the American Association of Engineering Societies centered on the utilization of engineers in the United States. Of those practicing engineers responding to the survey, 92 percent said that good communication between management and technical personnel is a key element in the work environment, but only 43 percent rated their current situations good in that respect. And although 87 percent thought it important that engineers participate in decisions, only 31 percent thought such a process characteristic of their own organizations.

Regarding work assignments, 80 percent thought that employers should be concerned with enlarging the individual engineer's work assignment, but only 31 percent found this to be the case with their present employers.

Engineers' work should be complex and challenging, said 79 percent, but only 38 percent rated the work in their organizations as meeting that requirement.

It would be interesting to obtain comparable data for engineers in other countries, but, interestingly, the likelihood of that is partially a function of the culture in question.

Over the past few decades, it has been the practice in Japan to take a basic technology developed in the United States and elsewhere and to design and manufacture products that are equal or superior to their competition. This strategy has been aided by a reward system that rates engineering research, design, development, manufacturing, and marketing as co-equals. Such has not been the case in the United States, where research engineers have traditionally been ranked at the "elite" end of the scale. On the other hand, the Japanese have recently enhanced their research activities so that they might leapfrog some of the product development in the United States and elsewhere. This new objective may require some modification of the traditional incentives and rewards for research in Japan as well as of its educational system.

While the individual chapters in this volume are unimpeachably authentic, they are intrinsically anecdotal in that they represent, usually, a single organization or one author's experience and opinion. Thus, the reader is cautioned against undue extrapolation. Just as what happens at IBM is not necessarily representative of other U.S. companies, so what occurs at NEC or Siemens cannot be construed to be universally true throughout industry in their respective countries.

The editor is indebted to Bruno O. Weinschel for his encouragement in putting this volume together, as well as his advice during its preparation. Thanks are also due George F. Watson of AT&T Bell Laboratories for his aid in editing the individual articles, and to Morris Khan and Nancy T. Hantman of the *Spectrum* staff for illustrations and for typing and correspondence, respectively.

DONALD CHRISTIANSEN

PROLOGUE

The end objectives of engineering excellence are universal, yet the way in which it is achieved differs in different parts of the world. Even so, it is possible that we all, despite cultural differences, can learn from one another.

I was fortunate to be able to visit Europe and the Far East recently, where I learned something of the different patterns of utilizing engineers in those regions. I was also fortunate to serve as a principal investigator, along with Russ Jones, vice president of academic affairs for Boston University, in a National Science Foundation study on engineering utilization. It was interesting to compare the results of the study, which covers 57 locations of 22 major high technology employers of engineers, with what I learned during my trips to Europe and Asia. The great emphasis in various European countries and in Japan on demanding excellence in all phases of manufacturing is worthy of note.

This starts with product research, product development, and development of manufacturing processes, and it continues through quality control methods of reliability assurance to the training of customer, service, and application engineers. These are all part of the concept of "total manufacturing engineering."

A common practice in Japan is to expose a new engineer to many details of the factory as well as of marketing, in a carefully planned program of rotation. The United States, on the other hand, stresses engineering excellence in the area of research and development. In Japan, industrial research is targeted to support a marketing decision.

While engineers in any culture may strive to enhance the competitiveness of their industry, they can usually only do this within technical areas. Engineers have practically no influence upon government policies that control the availability and cost of capital. The same applies to the value of a country's currency in the foreign exchange market. The macroeconomic policies are determined by political compromises controlling the philosophy of taxation and of the national banking system. Nevertheless, these factors are quite important.

The engineering community and its industrial management can improve performance by analyzing and adopting the best management policies of Japanese, U.S., and Western European industries.

Among the important factors having an impact on engineering excellence are education and skills updating, familiarity with market demands and constraints, the social ranking of various types of engineers, mobility and loyalty, salary, and research emphasis.

Because of worldwide interdependence, industrial corporations in the fast-moving fields of electrotechnology must show great adaptiveness to changing world conditions and technical developments. Along with greater emphasis on value-added products, this means that excellence in the general education of the work force and in the continuing education of technical professionals has become a condition of economic and national survival. All countries that expect to remain competitive must update the skills and knowledge base of the work force, especially of their engineers, as a matter of routine.

European engineers, as exemplified by those of Siemens and Philips, have a broader and more extensive technical education than their U.S. counterparts. Likewise, the Japanese blue-collar work force has a uniformly better education in the sciences and mathematics than their U.S. colleagues. This enables them, for example, to apply statistical quality control at the workbench.

Knowing the needs of the customer is considered vital by the Japanese and European engineers, while its importance is downplayed in the United States. Few U.S. engineering schools consider it important to turn out graduates who can solve problems of design for manufacturable products or services; emphasize reliability, maintainability, and cost-competitiveness in the international market; and, above all, satisfy the demands of the end user.

On the other hand, U.S. engineering education stresses engineering science at the expense of engineering and industrial practice. The U.S. educational system also neglects communications and human relations skills.

The quality of the nonprofessional technical work force in Japan is high because of excellent and uniform precollege education. In West Germany, the concept of apprenticeship is widely practiced, upgrading the skills of the blue-collar technical employee. Apprenticeship systems are rare in the United States, where the general level of precollege education is lower than in Western Europe or in Japan.

In both Japan and Western Europe, the continuing education of engineers is generally accepted as a responsibility of the industrial employer. Such responsibility is encouraged by the lower job mobility, especially in Japan, where lifelong employment is traditional within the major corporations. In the United States, the larger and more stable corporations have effective continuing education programs. General Electric, IBM, and AT&T are outstanding examples. However, because of the relatively high job mobility, many U.S. corporations, especially the smaller ones, and many in the defense industries, are reluctant to make significant investments in the continuing education of their technical professionals. In contrast, Siemens spends the equivalent of about five percent of its total payroll on training its more than 300,000 employees. An even greater proportion goes for the education of technical professionals.

There are notable differences in the functions and practice of R&D, too. Many U.S. R&D projects, lacking a focus on the needs of the market, are technology-driven, often in a quest for knowledge. Japanese developments, on the other hand—even those of a fundamental nature—are principally market-driven. European R&D falls somewhere between the two extremes.

In Western Europe, as in Japan, there is continuous emphasis on improvements in small steps in product quality and reduction in cost. Japanese R&D is carefully targeted. Product decisions are made by top management and only afterwards is the development of new supporting technologies scheduled. U.S. research is too often divorced from the needs of the marketplace. While this at times leads to major breakthroughs, a better balance between targeted and untargeted R&D appears to be more desirable. A figure of ten percent for blue-sky, or untargeted, research may be a better plan.

In contrast to the situation in the United States, the Japanese and Europeans highly value the importance of manufacturing. Their best engineers are guided into manufacturing engineering, while U.S. engineering graduates emulate the values of the engineering faculty, which is deeply rooted in basic untargeted research and too often lacking in links to current industrial and engineering practices. U.S. industry still pays manufacturing engineers less than R&D engineers.

In Japan there is no salary differentiation between engineers working in R&D, manufacturing, or marketing. Furthermore, there is a realization that every part of an engineering organization must be sensitive to the needs of the marketplace. Similar considerations apply to Siemens and Philips. For rapid and effective technology transfer, the development engineer must be closely coupled to the end users, and travel with the product downstream from R&D through manufacturing all the way to customer training and service.

He then returns to R&D as a more valuable member of the staff. In Japan, where the manufacturing divisions are profit centers, most of the engineers take part in marketing.

On the other hand, in the United States there is frequently a compartmentalization of product research, manufacturing process research, product design, process design, manufacturing engineering, quality assurance, application engineering, and service engineering. Occasionally, some companies form permanent teams from members of the various engineering departments to function throughout the life of the product. The result is quicker technology transfer, close coordination and quick reaction to problems of either the manufacturer or the user.

This volume, we believe, will yield useful lessons for readers regardless of their national culture.

The Japanese, for example, may well find it useful to increase somewhat the percentage of their R&D effort devoted to untargeted research.

For the United States, one message seems clear: education goals must be restructured. There must be greater emphasis on the continuing education of engineers so that the United States can preserve the value of a critical technical resource, its technical human capital, and extend the useful professional life of engineers.

A second message relates to the qualification of the U.S. engineering faculty; a minimum of two years of appropriate industrial experience is recommended. Engineering faculty must be more closely coupled to industrial problems, since most of their graduates work in industry and only a small portion end up in R&D.

Finally, a legislative initiative in the United States to grant a tax credit for the expenses of continuing education for sorely needed technical professionals would be a strong incentive to enhance the utilization of engineers. It would motivate industrial corporations and engineers alike to support programs that would maintain the competence of the engineering work force in the light of rapidly developing technologies.

On a global basis, it is important to encourage engineering excellence not only in basic research but also in innovation, development, and manufacturing engineering. In all of these phases, the technical professional must be closely coupled to the end user to ensure a high quality product or service with which the customer is completely satisfied.

Bruno O. Weinschel
President, Weinschel Engineering Co.
1986 IEEE President

Part I

The Importance of the Individual Engineer

The Contributors

Masaru Ibuka Honorary Chairman and Founder, Sony Corporation, Tokyo; Life Fellow, IEEE
Piet Kramer Senior Managing Director, Corporate Research, Philips International B.V., Eindhoven, The Netherlands
Dean O. Morton Executive Vice President and Chief Operating Officer, Hewlett-Packard Company, Palo Alto, California
Robert N. Noyce Vice Chairman of the Board, Intel Corporation, Santa Clara, California; Fellow, IEEE
Michael F. Wolff (discussant), Editor, *Research Management*; Contributing Editor, *IEEE Spectrum*; Member, IEEE

1/KEYS TO EFFECTIVE USE OF ENGINEERING MANPOWER

Masaru Ibuka

In 1946, I wrote a prospectus for a new company—a company that has since come to be called Sony Corporation. In the prospectus, I emphasized these points:

- We will create a corporation in which all people, particularly technical employees, are respected and are able to work to the best of their ability.
- We will not imitate the products of our competitors, but will try to create goods that have never existed in our market before.
- We will focus on the consumer market and apply the most advanced technology to the consumer products area.

As Sony's founder, I was able to shape the company spirit according to these principles. I feel that this spirit has been kept very much alive throughout Sony's growth. And I believe that this spirit has enabled our engineers to contribute more effectively to the company and has helped to make their work more fulfilling and rewarding.

I started in 1946 with capital of no more than $500 and with only 30 employees. The environment was a totally destroyed postwar Tokyo—one that I believe would be difficult for those who were not there to imagine. That first year, sales totaled $5000, and we ended up with a profit of $24—I was relieved that at least we had not lost money. By contrast, our 1985 annual sales figure was $6.7 billion. Through the years, our profits also increased. By 1985, profit had reached $630 million.

Unique products

Since we had decided at the outset not to imitate others, our products were naturally unique and were well accepted by customers.

We had no competition, and we were able to enjoy exclusive markets.

For our first unique product, we decided to produce tape recorders, which did not exist in Japan at the time. We had many obstacles to overcome. By regulation, we were not allowed to import parts or materials, so we had to make the magnetic tape for the recorders on our own. The manufacturing technology had originated in Germany and had been introduced in the United States after the war. But no documents were available to us, and we had to develop our own processes. Moreover, we could not even get the plastic material to make the tape base and were forced to use paper tape, coating it with a solution of magnetic powder using a brush.

Solving problems like this one was a real challenge, but we surmounted such difficulties one by one through our own efforts. The experience gave us the confidence to tackle the unknown in later years at Sony.

The next major project was producing the transistor radio. When I first heard of the invention of the transistor in an article in *Business Week* in 1948, I imagined it was something similar to the crystal detector and thought it would be unstable; I did not expect very much from it. But my ideas changed completely when I first came to the United States in 1952 and saw the alloy-junction and grown-junction types. I also learned from a friend in the States that Western Electric was going to release the patent.

At that time, two-thirds of the Sony employees (who then numbered 120) were college graduates hired for the tape recorder project. I didn't know how to make use of these highly educated people when the project was completed. It immediately occurred to me that development of the transistor was just the work to challenge these educated people, although the future of the transistor was not yet clear. Everybody thought that transistors would eventually be used in radios, but not one of the major companies ventured to produce such a radio because it would be extremely difficult to manufacture radio-frequency transistors.

Nevertheless, we acquired a license to manufacture the transistor from Western Electric. Western, out of concern for us, cautioned us not to get into the radio field.

The only consumer market for transistorized products at the time was for hearing aids. In Japan, this was a quite small market. So I thought, if we were going to manufacture transistors at all, we should aim at the radio market. Portable radios with vacuum tubes were coming into fashion in Japan in those days; it was considered a status symbol to have one. Clearly, a superior device to the vacuum tube was strongly needed, and we made the decision to go ahead and produce the transistor radio when the production yield for high-frequency transistors was only a few pieces out of a hundred.

I firmly believed that unless the yield was nil, it would be possible to raise it by finding, one by one, the obstructing factors and eliminating them. The production yield problem came up with all our major products later, but this first experience taught us not to be afraid of it. In fact, we have been happy to take on the challenge of products with low production yields.

We finally succeeded in manufacturing the transistor radio in 1954. We would have been the first company in the world to make the transistor radio if an American company called Regency had not done it six months before us. Nevertheless, our success in making the transistor radio in Japan taught Japanese manufacturers to take their domestic competitors seriously.

Since we were not able to manufacture the world's *first* transistor radio, we shifted our target to making the world's *smallest* transistor radio. We had to develop all the miniature parts ourselves. In doing so, we found that, to make something new for the domestic market, Japan depended on models from abroad—and indeed had depended on nonindigenous sources throughout its history. I believe that our attempt to develop smaller parts before any other country triggered the amazing growth of the electronic components industry in Japan.

Now for TV

Our next target was to transistorize the television receiver. Sony was probably the only company that had such a bold goal in 1962, because silicon transistors were still something special and cost more than $10 apiece. Even in the United States, it was unthinkable to use silicon transistors in a consumer TV set.

Fortunately or unfortunately, we did not have facilities to manufacture vacuum tubes and therefore decided not to use a single vacuum tube. We would build an all-transistor television. We made strenuous efforts to produce silicon transistors at a price suitable for the home market. But since it was technically difficult to produce a high-power silicon transistor, the biggest TV set we could successfully make was a micro TV with a 5-inch screen, in black and white, of course.

We never did expect that a small-screen TV would make a big, worldwide hit; many marketing survey experts in the United State had predicted that it would not sell. Nevertheless, the micro TV enjoyed a respectable success. It shares with the Walkman the distinction of being the only Sony products to be sold at a big premium on Fifth Avenue in New York City. More important, the micro TV inspired the trend to transistorize the television set around the world.

Our next challenge was to design a new color TV. Since we were latecomers in the field of color TV, we hesitated in adopting the RCA shadow-mask system that was then used throughout the world. We finally came up with a completely new system which we named the Trinitron system. Even though I was the president of the company, I became the project manager for developing, manufacturing, and merchandising the product, which turned out to be a valuable experience for me.

The Trinitron system had an impact on the shadow-mask system in general. Because of it, the shadow-mask was changed from a round hole configuration to a thin parallel slot configuration, and the gun alignment was changed from a three-gun delta arrangement to an in-line three-gun arrangement. The in-line guns had been invented by General Electric, but almost no one had produced them before us. Once we started using them, all shadow-mask manufacturers followed suit.

VTRs for the home

The videotape recorder, or VTR, opened a vast new market for Sony. In 1956, Ampex started to sell a four-head VTR for broadcast use. The product was a surprise to most of us. On hearing the news, the Sony development staff tried to duplicate the technology and soon came up with a working sample without having looked at a diagram or document on the Ampex machine.

From this experience, I learned how important it was to have a clear production goal. I learned that Sony was capable of achieving higher goals, of conquering very difficult technical problems, only if top management gives the targets. We could have produced the VTR earlier if we had set up the target earlier.

My colleagues at Sony insisted that we make and sell broadcast VTRs similar to Ampex's. They thought that we would thereby make quick profits. I strongly opposed the idea and blocked action on it. I directed our engineers to create another VTR format for home use.

At first, the Ampex black-and-white video recorder cost $50,000. In 1962, Sony succeeded in selling institutional video machines at the price of $5,000. The first customers were an American horse-racing company, an American airline company, and Pickering, the x-ray company.

In 1964, we succeeded in mass-producing a video recorder at less than $1,000—a somewhat desperate goal we had set for ourselves. This was the breakthrough into the home and educational markets. Finally, in 1970, we started marketing a cartridge-type video recorder, and this kicked off the age of small cassette VTRs.

Thus, a whole new industry for videocassette recorders was created. The size of the total VTR market grew to become comparable to the TV market. A tremendous amount of technical innovation made it possible for this great change to happen in such a short time. Japan is often accused of being a free rider, of making no contribution to invention and innovation. But, in fact, Japanese manufacturers have produced a great deal of new technology.

An interesting irony is that Sony's accumulation of technology in the consumer VTR market made it possible for us to enter the market for professional VTR equipment. Right now, most of the broadcast VTRs used throughout the world are Sony-made.

A different path

In the ordinary course of development, basic invention comes first. The invention is developed for application and a new product is born as a result. At Sony, we have demonstrated a new path for making a product: specific targets are set first, and the technology for achieving the target is developed later through research and development.

I have always tried to develop a product that creates a new field. I strongly believe in using high technology not yet available in the marketplace or available only in the aerospace or defense fields.

Incidentally, I have always disagreed with the prevalent view that the absence of a big defense industry in Japan delays the development of new science and technology. I can assure you that we strive to use newly discovered principles and newly invented technology in our consumer products.

Since we do not use easily available technology at Sony, our research requires considerable investment and manpower. Our ratio of research and development expense to sales has been around eight percent or more in the past few years.

People who love to work

Finding the best people to work on Sony's R&D isn't easy. My criterion for hiring has been to hire those who love to work. It is of secondary importance to me how much technical knowledge or skill a person possesses. Difficult projects give employees the best training and experience; what they have in the beginning does not matter much. I attach more importance to an employee's eagerness to work and ability to harmonize the conflicting interests of colleagues.

Since Sony always ventures into completely new areas of exploration, it is difficult to find specialists in these areas. Not until one has worked on a project for a period of time can one realize what kind of research is necessary to accomplish a goal. It is common for our engineers to engage in a great deal of research work on their own, in subjects far removed from their own specialties.

Humans cannot produce good results merely from compulsion or a sense of obligation. Results can be achieved only through their ambition to create something new and their persistence in addressing the task.

In order to achieve these goals, we don't have clearcut divisions of labor regarding who is in charge of R&D, or production, or quality control, as they are commonly defined. In my own view, those who are engaged in a project should oversee the whole process from design through sample-making, production, quality control, and marketing the product. It would be ideal if these same people could sell the product, talk to the consumers and even provide post-sales service. Solving quality control problems should not be the responsibility only of the handful of people who are in charge. Rather, the conscious dedication of employees engaged in each step of the production process counts towards achieving good quality control.

Such results are encouraged all the more by a characteristic feature of our company: there is no clear difference in treatment between blue-collar workers and white-collar workers. Top management strives to create an environment in which each Sony employee—from the highest levels to the lowest—feels rewarded and filled with the joy of working. The larger our organization becomes, the more important it will be for top management to ensure this environment.

Education and training are also very important in bringing prosperity to a corporation. They are something that no top manager should neglect. They also happen to be the subject of my greatest interest in recent years.

Meanwhile, as we put computers and robots into our plants, it becomes clear to me that the twenty-first century will not be a mere linear extension of the present. Building warm human relationships in the potentially dehumanized environment of the private corporation will become ever more important.

2/DEVELOPING AND MOTIVATING ENGINEERING PERSONNEL

Piet Kramer

To speak about "development" and "motivation" is a risky venture. Even if the concepts are thought of in a general way, they mean different things to different people in different places. Not being a social scientist, I feel somewhat diffident about expressing my opinions on these subjects.

However, a teacher once said to me, "If something is so complicated and uncertain that nobody can say anything about it that makes sense, why don't you?" I have taken this advice to heart.

Certainly, I know that engineers can be developed and motivated and that we have had some success in doing so at Philips. Here, I will discuss five aspects of development and motivation. I will begin by sketching a profile of the engineer in Western Europe. Then I will ask the question, "What motivates this engineer?"—and, I trust, answer it. To help people understand engineers at Philips in the Netherlands in particular, I will describe some characteristics of our organization and of the European welfare state. I will examine pertinent changes in the organization and in society, and finally I will focus on the development of the employee.

Defining terms

My company, with 344,000 employees, is one of the biggest in the world. We have industrial activities in 47 countries and a commercial presence in many others. We have nine divisions whose products range from lamps to electronic microscopes, from shavers to radar systems, from compact disks to telephone exchanges. Some of our products are mass-produced for anonymous consumers; others are complex one-of-a-kind systems developed in close contact with the client. Some of our products are simple and straightforward, whereas others are technologically sophisticated.

Worldwide, we have 14,000 engineers working in process development, 19,000 in product development, and 4,200 in research. About 30 percent of these employees have an advanced university degree.

I am giving these facts to point out that each division and product line involves a different culture. An engineer working in the lighting industry in India has an orientation different from that of an engineer developing medical equipment in the Netherlands. Motivation and development needs differ from one person to the next, are influenced by the situation, and are conditioned by the culture. I am therefore confining my comments to engineers in Philips' Western European factories.

For the sake of precision, let me define what I mean by the words "motivation," "development," and "engineer." An engineer is an employee with four to six years' tertiary education at a university or an equivalent institute. By development, I mean further university education, postgraduate courses, and training programs.

Motivation is more ambiguous, and even when people agree on the meaning of the term, they find it difficult to measure. As an approximation, I will define or circumscribe motivation as something that stimulates people in their work, something they desire and expect.

But I must add some qualifications to this definition. Implicit in it is an acceptance that what employees say is what they mean, but we are all well aware that there are pitfalls in taking expression of wants or requirements at face value. Often people will say what they think the questioner wants to hear or what they think conforms to the norm, while deep in their hearts they want something totally different.

Moreover, people often don't really know themselves or can't put themselves in a hypothetical situation—they can't imagine what it's like to be unemployed or retired.

Most important, either of two psychological mechanisms may be at work when people try to express what they want from their jobs. On the one hand, there is the "coping mechanism": placed in a situation that does not fit with what was originally wanted, an engineer may adjust values, adapt wants. On the other hand, there is what I call the "progressing expectations mechanism": people want ever more, and this inevitably leads to disappointment; once a need has been filled, one hardly has time to be happy before new desires appear.

The Western European engineer

I will describe the engineer in Western Europe by highlighting differences with the engineer in the United States. My views are based on what I have seen during several visits to the United States and what I have learned from others with whom I have discussed the differences.

In general, the Western European engineer is highly educated—probably more highly than the average U.S. engineer, although the top universities of the United States can hardly be beaten. The program of study in European universities is probably broader than that of the average U.S. university. The European engineer is less specialized, and is therefore more flexible and able to change to a neighboring discipline.

I believe that the U.S. engineer is a bit more aggressive and more individualistic and keeps his own career foremost in his mind. When the U.S. engineer sees better opportunities, he does not hesitate to change employers. In Europe, engineers are more rigid in this respect and more loyal to their employer. However, there is a trend these days in Europe toward a more dynamic relationship with an employer in which loyalty to the engineering profession is more important than loyalty to the company.

The general characteristics of an engineer are identical in both the United States and Europe. The engineer is studious and creative. The engineer is an achiever—he wants to see results. The engineer has a need to be needed and wants to be part of the organization.

What motivates the engineer?

Cynics say that labor ethics have disappeared, that today's workers, especially the younger ones, are not interested in their work, that they have no motivation. This is nonsense!

Traditional norms and values may have disappeared, but new norms and values have developed. Traditional work ethics may have disappeared, but new ethics have emerged.

In the past, workers were primarily interested in high income and a career. They looked on working as a moral obligation. (This is what Weber's classical sociology called "the Protestant ethic.") But in a study my company made of changing attitudes toward work, we found that new priorities have arisen. People now give greater importance to developing themselves and realizing their full potential, to being challenged, to feelings of being useful and of belonging to an organization.

We at Philips believe that knowing and understanding what motivates workers is of great importance to the organization. We believe that the modern organization should adapt to the new attitudes; making use of them will benefit both the company and the employees.

This applies to workers in general, but what about engineers in particular? An engineer attaches high value to having an assignment of his own and having autonomy over it. Of course, complete autonomy cannot be allowed in a production organization. However, given a certain job, the engineer should be left as free as possible to plan, organize, and carry out the task. In product development, for example, employees should be allowed to spend a certain amount of their time—say 15 percent—on "free" work on activities of their own choosing related to the product. They might work on alternative approaches or on new versions of the product, but they should also read technical papers.

Engineers should feel that when they do first-class work, management will recognize it and show appreciation for it. And management should support first-class work by providing engineers with modern equipment of a high standard. If researchers and developers can't use the latest apparatus, they might look for a job in another company. Remember that loyalty to the profession is becoming more important than loyalty to the company.

Whenever possible, the worker should have a task with which he can identify personally. He should also be rewarded for it personally—not necessarily in money. Recognition from the boss or from peers might actually be more important.

In complex organizations, engineers will be more specialized. They should still see their work in relation to others, however. They should get information to put their work in perspective from company newsletters and other publications, from group discussion among peers, and from regular meetings with people higher up. This kind of information is not only needed for organizing the work, but is also a prerequisite to motivating the engineer.

Many workers are inspired by knowing that their company or business unit is successful in the marketplace. They see their personal efforts translated into commercial results. They see such success as a sign that they have contributed to fulfilling a real need and that fruits of their efforts can compete with those of others. Success appeals to what the sociologist Veblen called "an instinct to be efficient and effective."

On the other hand, it can be demotivating to some engineers when their company is not doing too well. With them, there is danger of a vicious circle: just as success can create success, so failure can lead to new failures. However, for employees with a fighting spirit, a survival situation challenges them—they will do all they can to surmount the difficulties. (Incidentally, I envy those Japanese companies that, at a pinnacle of success, can tell employees that they still have to do their utmost for survival of the corporation.)

Perhaps the strongest motivation can come from an inspiring, challenging project. For Philips, the megabit memory development and the compact disk effort were such projects. Employees felt a new atmosphere of entrepreneurism and élan.

Of course, not all tasks or products can inspire, and what was once exciting can become a routine affair. Motivating engineers in their midlife, those who are at the peak of a perhaps mediocre career, who have a regular, routine job—this can be a real problem. Sometimes job rotation or job enlargement (for instance, adding some managerial work) can help. Or perhaps the engineer can be sent to a refresher course or a general business course. This will give a feeling that the company still finds it worthwhile to invest in him.

Rewards in the organization

Coming from the Netherlands, I am always struck by the importance attached to making money in the United States. In the Netherlands, financial rewards are only part of the expected remuneration.

As I noted, recognition by the peer group and by the hierarchy is important in the Netherlands. Awards, however, are not. To be proud of an achievement award is considered childish. Quite the opposite is true in the United States, of course, and to a lesser extent in France and the United Kingdom. But Dutch attitudes may be changing. Recently, within Philips, a quality award and an energy award were established and were well received. Nevertheless, they still seem strange in our corporate environment.

A change in the status of engineers and in motivating forces may be brewing from the enormous increase in the number of students in the last decades. Becoming an engineer no longer implies entry into an elite group; in a society in which an academic degree is no longer exceptional, the demands and expectations of the graduate have to be modified.

Recently, we engaged a great number of qualified young engineers; we can offer them high standing and interesting work. However, whereas in the past one out of six of the newcomers would rise to a higher management position, this can no longer be the case. This decrease in upward mobility will inevitably lead to tensions in the organization in the future.

The problem will be mitigated somewhat by the fact that many engineers are not at all interested in a management post. For some of them a new rank in the organization may be created—that of scientific advisor, without managerial responsibility. In this so-called dual ladder system, they will still be able to advance and find higher rewards.

Another mitigating factor is the trend to a flatter organization, one with more middle management. I will elaborate on this development later.

Certainly, I believe that a company should be open with new employees about the potential for upward movement to avoid creating unrealistic expectations. This practice will help to prevent conflict and demotivation in the future. At the same time, a company should highlight its attractive characteristics. At Philips, we can tell employment candidates that we are an international, diversified company that offers considerable freedom to its employees and is active in fascinating, advanced technologies.

Such characteristics can be powerful motivators. As I have noted, for many engineers it is highly challenging to do high technology work in one of the world's top laboratories with the best equipment. Perhaps even more appealing to young engineers at Philips is the freedom the company offers to its personnel. Traditionally, we have been organized around our people, not the other way around. To outsiders, our loose organization sometimes seems unworkable. In fact, a leading U.S. consultant once characterized Philips as "a gathering of good, willing people working in a random fashion."

Our lack of structure and of strictly defined responsibilities may appear to be a weakness. But it often turns out to be one of our strengths. To many of our best employees, this free-for-all climate is stimulating.

And many young engineers value the opportunity to go abroad or to have international contacts that a worldwide organization affords. Perhaps even more value the opportunity to work in a great variety of jobs that our diversified company offers.

Working in a changing organization

The electronics industry is nothing if not dynamic. The market situation, the products, and the production processes of tomorrow will differ from those of today just as surely as today's differ from yesterday's. And this dynamism will continue to affect the organization, the employees, and the quality of work.

At the same time, the employee will continue to change. The

engineer will become even more highly educated and more emancipated, and his attitude toward work will continue to change. And more and more, this engineer will be a "she," not a "he."

How can we continue to motivate and develop engineers in this often tumultuous environment? Without indulging in speculation about the future, I'll examine trends in the electronics industry as we see them at Philips, and how they affect motivation and development.

One trend is the growing importance of software and computer aids. Newly hired engineers will generally be better trained in these technologies than older employees. If this does not result in cultural shock, it will certainly lead to tensions within the organization. Managers of the old style may find it difficult to understand and manage the new generation of software engineers. Since software engineers are in short supply, they may get higher salaries than the older engineers—and that understandably may be demotivating for the latter.

And what about those same software engineers? After sitting before a computer terminal for 20 years, will they still find their work inspiring? Large-scale boredom is a problem that might arise, and we should be prepared for it.

Another trend is the increasing complexity in products and production processes. This trend is pushing responsibility to lower levels in the hierarchy, to the specialists. It fits in with the demands of engineers for more autonomy and responsibility, but it also creates coordination problems, since responsibility is becoming so much more spread out, and at lower levels of authority, than it used to be.

Also fitting engineers' preferences—and also creating coordination problems—is the trend to small-scale production and business units under the umbrella of a large-scale company. The researcher, the development engineer, the product designer, and the marketer must work more closely together. In other words, with greater autonomy comes the need for greater discipline, for more structured problem solving, and for greater harmony and cooperation among people who traditionally have held diverse viewpoints.

Paradoxically, some engineers look on greater local autonomy as a loss of personal autonomy. Certainly, there will be no place for hobbyism and the quick-fix solutions that some engineers are so fond of.

Professional development

Development is essential for motivating engineers and for coping with the problems that come with changing situations and attitudes.

In a dynamic environment, continuing education is a must. Since our engineers at Philips have gotten a broad education rather than a specialized one, they must learn specialties on the job.

Philips gives ample opportunity for that. Discussions with colleagues, contacts with research people, reading literature, and attending conferences are all encouraged.

Formal courses are organized as the need arises. For example, we've instituted a training program in information science. In the coming four to five years, 20,000 employees in the Netherlands (out of a total of 78,000) will take at least one of the courses. For product development engineers, we've set up a training program on methodology of design and innovation. And, of course, we sponsor many courses for general orientation and management training.

As in most companies, supervisors are often reluctant to give their best engineers time off for courses. Their efforts are needed on the job. That's true, but the engineer and his work will benefit not just from the course work but also from the refreshment that a break from work affords.

At Philips, the responsibility for education and training is shared by the central organization and the separate divisions. The central organization, for example, operates the information science program and the product development course. We also make use of outside education and training institutes.

We keep in mind that engineers are the true engine of our company. We know that it is up to us in management to pay ever greater attention to the proper use and maintenance of this wonderful engine.

3/HOW TO ATTRACT AND RETAIN INNOVATORS

Dean O. Morton

When I joined Hewlett-Packard in 1960, the company had a great setup for innovators. Everybody worked on a single product, a voltmeter or counter or oscilloscope. Product development teams were small and their charters were clear and relatively independent. The teams were part of semiautonomous profit centers that were managed like independent businesses by engineers—electrical engineers at that.

This scheme worked well. Communication among the team members was easy and they had almost no ambiguity to deal with. The objective was to make contributions, to push technology—largely proprietary—along a natural trajectory, and to develop products to be sold to and used by other engineers. There weren't a lot of market research people in the act.

One of the nicest things, looking back on those days, was that the competitive environment and the desirability of our own products allowed pricing by formula. It was a pretty simple formula: you just took the manufacturing costs and multiplied them by 3 or 4 to get the price. There were no elaborate competitive analyses or evaluations, no questions about family pricing strategies.

Today things are more complicated. Innovation is as important as ever, but the conditions have changed. We still have project teams, but they are not so small anymore and they don't enjoy the freedom of stand-alone product charters. Technology and customer needs have moved us into the system business. Our focus has shifted from products to solutions. Whatever is done in one group is linked and interdependent with a lot of other groups. Cross-functional dependency is a fact of life: marketing and manufacturing have a lot to contribute to successful product programs, and collaboration is just part of the job.

Even though H-P is a hundred times larger than it was when I joined, it is more tightly integrated and has more mutual dependencies than 25 years ago. We want to operate as one company and the reorganization that we put in place about a year-and-a-half ago emphasizes that. These things obviously change the way we manage and the way people spend their time.

One thing hasn't changed

But one thing hasn't changed, and that is our emphasis on attracting and retaining innovative people. I think H-P was fortunate in having founders who recognized and respected the value of the individual. This heritage finds many expressions in our company, but it really begins with hiring the very best people we can. I don't know of any better formula for success. The accumulated impact of bringing top-flight people into the company year after year is hard to beat. I think it is really the continuity of our college and university hiring programs that is at the center of the success of the company and accounts for our reputation for hiring people who prove to be innovators.

We have recruited at the top 100 schools in the United States. We go back every year with teams that build relationships with the schools just as our account management teams do with major customers. The teams compete with each other for results. They get to know the faculty and they try to identify the most promising students well before the on-campus interview date.

In 1985, which wasn't the best of times for our industry or for Hewlett-Packard, we still hired 1,300 people from colleges and universities in the United States. Most were technical people. About three-fourths had B.S.-level degrees; the others had graduate degrees of one kind or another. Most of the people we hired in 1985 were electrical engineers or computer science people. A fair number were mechanical engineers and a few were MBAs.

We looked for people with problem-solving ability, not just knowledge. Each of our recruiters has favorite questions aimed at evaluating innovative skills. They ask candidates such things as:

- What was your part in that endeavor?
- How was the outcome improved by your contribution?
- What was special about that experience?

They try to understand the motivation of people and how they approach problems. One of recruiter Barney Oliver's favorite questions—one that I have only recently figured the answer for—is "Why are manhole covers round?" Questions like these help us to find the kind of people with the innovative mindset and the drive fundamental to our continued success.

We look for flexibility and openness and, increasingly, we look for the ability to work with others. I said that the process of innovation is changing and that the company has changed as we move toward larger systems, more interdependency, and more teamwork. The ability to join in is clearly an important requirement today for everybody.

We hire people who have also done well in school. I don't think grades are everything, but your chances of being hired by Hewlett-Packard are best if you're in the top 10 or 15 percent of your class. Over half of the people that we hire have grade-point averages above 3.5.

I think practices like this become self-reinforcing as established employees become the ones to hire the next generation of employees, and successful innovators attract others of similar potential. And as a company, we try hard to make clear our commitment to innovation and to articulate our understanding of the role it plays in our performance. We regularly spend over ten percent of revenue each year on research and development. H-P's R&D currently ranks in the top three of industrial electronic programs in the United States.

New products are the key to our growth. If H-P has a central value, it's the importance attached to new products and the contribution that new products make to the growth and the prosperity of the company. In 1985, 75 percent of our orders were for products that were less than four years old. We try to make sure that engineers are recognized and rewarded for the fundamental role that they play in this.

Engineering at Hewlett-Packard

What do engineers do at H-P? About 45 percent are either doing design or are involved in design and engineering management. Of them, about 15 percent are in research and advanced development of processes, materials, and new products; that is, they are a part of what we call Hewlett-Packard Laboratories. Twenty-seven percent of our engineers are in marketing, 24 percent are in manufacturing, and another 3.5 percent are in various other departments: finance, administration, and some even in corporate management.

How do they spend their time? In much the same way described in the American Association of Engineering Societies study on utilization of time. Most of their time goes toward documentation, planning, and communication. When I think about managing our engineering resources and improving productivity, it seems clear that this large element of nondesign, nontest, nonverification activities needs special focus and attention. I think the time distribution of engineering activities is greatly different from 1960. In fact, it is possible today to succeed in this business without even being an engineer, because of growing software content and increasing numbers of nontechnical customers.

Probably the most important trend in the practice of engineering at Hewlett-Packard is the growing role of marketing and manufacturing. They play a key role in making sure that our engineering resources are appropriately directed. This doesn't mean that engineers don't have the full responsibility for understanding customer needs and for applying technology to the fulfillment of those needs. Nevertheless, it's a very complicated world, and understanding what goes on in the world in terms of requirements and applications requires lots of people. This applies to manufacturing, too; lowering cost and raising quality in manufacturing and design for manufacturability are key competitive elements in the survival of any of us.

Of course, it's not enough to attract innovators—we have to retain them once they are hired. H-P does pretty well in this respect, but we lose some good people too. We are never happy about that, although many of the people who leave at least turn out to be good H-P customers, and some of them come back in time. We do take an ambivalent pride in being the kind of company that recruiters look to when they are trying to find good people. Nevertheless, overall our attrition rate for engineers is quite low by industry standards. It runs less than six percent a year, and in fact it's been coming down over time.

I think our ability to retain good people, especially innovators, comes from several factors. You have to pay competitively, of course, and you have to recognize contributions and provide opportunity for professional growth and career development. Beyond that, and I think more important than that, innovators like H-P because they can do more with us than they can with many other employers. We are an engineering-oriented company. Innovators find stimulating peers, people they can relate to and talk with in their jobs. We provide ample resources—and it takes a lot of resources today to tap creativity—so that our engineers can optimize their productivity and move technology to the marketplace quickly.

And finally, we do encourage innovation. I can't think of any better example of that than our Spectrum program in which H-P recently made a major commitment to reduced instruction set computing, or RISC, concepts in processors for its next generation of products. This is an unequivocal commitment; we have no standby plan. It's referred to in the press as a "bet your company" kind of program, or H-P's Manhattan Project, and perhaps it is. But we decided to abandon a more conventional and perhaps safer architectural approach because RISC is innovative and offers the potential for a significant competitive advantage. It also provides us the opportunity to direct R&D resources by means of this architectural approach into a broad family that could fill the needs of both our technical and commercial customers.

That kind of visible commitment to innovation and well-informed risk taking does more than anything to make and keep H-P the kind of place where innovators want to work.

4/THE IMPORTANCE OF THE INDIVIDUAL'S CONTRIBUTION TO INDUSTRIAL PROGRESS

Robert N. Noyce

The notion of the rugged individualist overcoming all obstacles has characterized U.S. cultural history from its beginning. The same spirit of innovation that brought the colonists to America and later spurred the settlement of the West by the pioneers also provided the foundation upon which so many of our key industries were built.

When many of us followed Horace Greeley's advice, "Go West, young man," a few decades ago and came to this valley—Silicon Valley—none of us thought that our research would lead to the multibillion dollar integrated circuit industry we are part of today.

But, successful as it has been, that industry is now under siege. Driven in large part by foreign competition, difficult market conditions are forcing us to approach our business differently.

Innovative minds function in a framework, and the framework of our industry is threatened by the current economic and industrial climate. That framework for innovation and the threat facing it are what I will focus on.

All great ideas start at the individual level; I don't think anyone would contradict that. The essence of creativity is the ability of the individual to break through conceptual barriers and open the eyes of the world to previously unexplored opportunities. We certainly don't have to worry about a shortage of ideas; that's not the issue.

What is important is that we as an industry, as a nation, and as a society continue to provide a nurturing environment—one conducive to innovation, where the seeds of creative thought can grow. That's where the issues of economics and industrial competitiveness come into play. Our semiconductor industry, in particular, has faced troubled times over the past year and a half. It's hard to think about extending frontiers when many companies are struggling to stay out of the red.

You may be familiar with Maslow's hierarchy of needs (Abraham H. Maslow, "A Theory of Human Motivation," *Psychological Review*, vol. 50, pp. 370–396, 1943). Starting at the physiological level of survival, Maslow's hierarchical ladder stretches upward to the top rung, self-actualization—the rung of the creative mind. Maslow maintains that no individual or group can exist on the upper levels of the ladder until each of the needs of the lower levels has been met.

In other words, you don't worry about self-actualization when you're starving to death. Applying Maslow's model to the plight of our industry today, we can see that it's difficult to provide an environment conducive to innovation when the issue of keeping our companies operating profitably is staring us in the face.

Without robust economic growth, there isn't enough of what I call "slack" in the system. This slack—economic solvency, if you prefer—is what allows corporations to experiment, to take risks. Individuals as entrepreneurs or as part of a corporate structure are less likely to innovate if they cannot pursue ideas for the sake of curiosity, if they must be ever-mindful of economic survival.

I'm reminded of some of my colleagues in the past who would put off a critical experiment as long as possible because it might disprove a thesis they had proposed. They feared failure, even though they had the full support of management behind them. And when slack is removed, failures become critical because they can affect the solvency of the entire company. It's extremely difficult for creative minds to function when they're under pressure not to fail, when their backs are against the wall.

Freedom to think

We can't stand over people with a whip and tell them, "Create, or else." It just won't work. The process of innovation depends on our ability to give our innovative people breathing room and leave them alone so that their ideas can eventually bear fruit.

During the first few years at Intel Corporation, for example, we would budget a significant amount of money for new projects. We didn't always know what those projects would be. We just knew that we had good minds in our organization. We had to provide the money and the environment so that those minds would have the freedom to think and create.

That's how Dov Frohman invented the erasable, programmable read-only memory, or EPROM, at Intel, and how Ted Hoff developed the microprocessor. When they stepped forward with their ideas, it made sense to give them some room to experiment. There was no guarantee that their ideas would work. But we felt that we had to take a risk and allow them to succeed—or to fail.

As we all know, they didn't fail. Their ideas changed our industry, and with it, the way people the world over live today.

But those two devices—the EPROM and the microprocessor—appeared in 1971. Things were much different then. There was a general feeling of optimism then about the economic future of the United States. After all, we had put a man on the moon just two years before. We led in almost all key areas of innovation, and our position in the world economy was not seriously challenged. There didn't seem to be any limits on our horizon.

Now, 15 years later, we look at our industrial future with a certain pessimism. Our slack, our margin for error, has largely disappeared. Increasingly, we're finding ourselves playing things closer to the vest, not taking the kinds of risks that characterize the big breakthroughs.

I have a patch on my ski jacket that says, "No guts, no glory." And there's a lot of truth in that. You can't achieve greatness barreling down the bunny slope. As our economy tightens and our risks increase, we have put serious constraints on our ability to innovate, both on the individual level and on the corporate level.

An environment filled with pessimism breeds disorder. During good times, the pie is big enough that everyone is assured of a substantial piece of it. But during bad times, groups become more protectionist and less outward looking. They become concerned with only the issues that affect them directly. Many voices clamor for a piece of the smaller pie. It becomes harder to focus on long-term global issues, and everyone suffers as a consequence.

The right conditions at the right time

For times to be good—for an environment in which individuals and institutions have the slack to innovate—a lot of conditions must be precisely right at the same time. All the support was in place, for example, when the Japanese decided in 1975 to launch their attack on the market for dynamic random-access memories, or DRAMs. The Japanese did so because, looking at the factors in their political and economic environment, they perceived an opportunity and they believed they could win. That's a very important point: the partnership between government and industry was there to further advances in very large scale integration and to provide investment capital within a supportive framework. The partnership enabled the Japanese to take over the DRAM market.

On the other side of the coin, U.S. suppliers of DRAMs are

dropping out without a fight. The philosophy seems to be, "If you don't think you can win, then you can't." The lack of profits and investment capital to improve our manufacturing techniques, the strong dollar—these have combined to create an uncharacteristic pessimism about our industrial competitiveness that undermines the risk-taking process.

We must always find a way to let our thinkers think. Support of fundamental research has been the springboard to innovative breakthroughs. We must find a way to continue such support despite the current realities. We can't let what's happened at places like Stanford Research Institute continue. For almost 40 years, the people there have devoted themselves to basic research, and they've engineered all sorts of breakthroughs. But they, too, are being forced to become more market-driven, to look for economic return from their efforts. We have to do what is necessary to keep think tanks like SRI alive.

Leading-edge contributors in science and high technology pursue ideas for the ideas' sake, rather than for economic gain. Twenty-five years ago, who could have predicted that the unraveling of the double helix would lead to a new industry? That search was driven by curiosity rather than by economic motives. And in 1947, when Bell Laboratories announced the transistor, could anyone have said that semiconductors would blossom into a multibillion dollar business? (At that time, the $500 million vacuum tube industry seemed like big business.) Who would have thought that semiconductors would spawn the personal computer industry? We certainly didn't approach our research with that kind of thinking.

Turning the tide

Thus far, I've painted a grim picture. But there are answers, things we can do to turn the tide. A little over a year ago, I participated in a study undertaken by the President's Commission on Industrial Competitiveness, headed by John Young of Hewlett-Packard. The commission's report set forth a number of recommendations.

First among them, the United States must return to its fundamental values—the old "saving for a rainy day" concept. We have become a nation of spenders. One look at the budget and trade deficit tells that story. It's axiomatic that if we consume more than we produce, we must import. And that strengthens other economies, not ours. That removes the slack and ultimately reduces the importance of the individual as an innovator. If we lose our lead in innovation—a lead that has always fed on the notion of Yankee ingenuity—we're really in trouble. We have to overcome the buy now, pay later syndrome, or we'll be paying through the teeth for a long time to come.

Compare the consumer with a full year's salary in the bank with one who's in debt and buying everything on credit. The one who saves has a margin for error. The one who spends too much has no slack. Yet debt is rising. In December 1985, consumer spending was up two percent, the biggest monthly rise in more than a decade, according to a U.S. Department of Commerce report. The report went on to say that the debt created by our personal consumption had drained savings to the lowest levels since the Korean War. Shortage of savings has created a greater demand for the dollars that are available. This demand drives the value of the dollar up, forcing U.S. industry into a noncompetitive position.

We need to produce, to create wealth. We find ourselves in a domestic economy more involved with manipulating existing wealth than creating new wealth—the service economy syndrome. It is the ability to create wealth through innovative ideas that has given us the high standard of living we enjoy in the United States. To maintain that standard, we must concentrate on producing more and consuming less as a society.

With the U.S. government as a role model, it will be hard for our society and our institutions to make such a change. Our government is hundreds of billions of dollars in the hole. Government must set the tone and take a more active part in protecting our economic interests. Federal deficit spending must be curtailed, even if it means imposing additional taxes.

We need tax reform. We need a sharply focused trade policy. And we need governmental encouragement of civilian research and development.

Let's look at our tax structure first. Economists usually agree that savings can be increased by taxing consumption and exempting savings. But legislation has been moving in the opposite direction, as taxes are shifted from individuals who spend to corporations that save most of their income. This doesn't bode well for the efforts of industry to put money into new plants and equipment. The system is not allowing us to grow.

Trade policy is next. Our government has yet to focus on trade as a national priority. Even though 15 to 25 percent of the manufactured goods we consume are now produced abroad, the perception persists that we are still a domestic economy. The Young Commission recommended that a cabinet-level Department of Trade be formed to meet this issue head-on. Without a national focus, it will be impossible for us to sort through the many problems that plague us on this front.

About R&D: even though most of our progress in productivity has come from advances in science and technology, we are now falling behind our trading partners in this crucial area. As more of our national research is devoted to defense—with its vaunted, but questionable, fallout benefits on commercial applications—our civilian research and development efforts are being shortchanged. The Young Commission recommended that a Department of Science and Technology be formed to assure that our nondefense R&D is coordinated and supported in the pursuit of our national goal of industrial competitiveness.

In the same boat

We must invest in the education of our present and future work forces. We must help workers adapt to the technological changes they will see in their careers as they grow older, and as they move from one part of the country to another.

The division between management and labor must disappear. We're all in the same boat. Traditional strife must give way to cooperation; without it, we will be unable to advance.

The issue of educating our youth for productive careers has been widely discussed. Suffice to say that the issue is still with us, and that the investment in youth through our universities is critical to our continued well-being and to the existence of the economic surplus necessary to foster creativity.

We're also going to have to learn how to manufacture better. The Pacific Rim nations have proved emphatically that they can make quality products at a lower cost than we can. The gap in labor costs is a big enough hurdle to clear without having to deal with all the other adverse factors I've mentioned.

We've gotten sloppy over the years because we've never had to worry about foreign competition before. It's time to turn our thinking to manufacturing and process technology. Process engineering must be elevated to the same exalted status that we have bestowed on product design. We also have to work smarter and get more out of the facilities we have on-line today. I know that we've made a major effort at Intel in the last year to turn our manufacturing performance around, and we're starting to see some real results. It can be done.

Finally, our corporations must cooperate with each other to

increase our national productivity. At a time when competition is global and each nation seeks to raise its productivity, we must recognize that the threat to competition lies not in cooperation between companies, but rather in the lack of it. By pooling the knowledge and resources of our leading corporations, America can become more competitive.

Back to the individual

It's ironic to talk so much about governments, global policies, and corporate cooperation when our concern here is with individual contributions. But unless issues about these large entities become resolved, we'll never be able to regain the slack that is so necessary for the individual creative mind to flourish.

I believe that the mindset to innovate and take chances is still intact; of course, it's imperative that we maintain it. Of greater concern is that our economy and industries can be successful enough that slack can be regained and the climate for innovation can be affordable once more.

We in the United States can no longer win simply by showing up. But we can, through a cohesive effort, put ourselves in a position to win. That's all you can ask for. Given that environment, I have no doubt that the individual in our society will continue to contribute and make those quantum leaps forward that have characterized the success of American industry for many years.

5/MANAGING HUMAN RESOURCES

Michael F. Wolff

More than 15 years ago, an editorial in *Innovation* magazine (where I was then working) decried the latest round of cutbacks in the aerospace and electronics industries. Noting that "since the mid-1950s, glamorized cycles of hiring engineers and technicians seem to be followed inexorably by cutbacks a few years later," *Innovation* wondered why industries that boasted of being in the forefront of technological innovation couldn't do better than treat their human resources as replaceable/disposable objects.

One reader replied: "A manager who was asked 'How many people work for you?' looked about his vast engineering office and said 'Oh, about $1\frac{1}{2}$ acres.' "

The previous chapters demonstrate that nurturing technical people is as important an issue today as 15 and even 40 years ago when the so-called human potential movement emerged. In the United States, Europe, and Japan, enlightened technology managers recognize the urgency of creating environments where individuals can be their most productive and creative. Indeed, with science and technology research results disseminated as quickly and widely as they are, there is the growing realization that the key to maintaining a competitive edge in world markets may well lie with the management of human resources.

The big questions, of course, are how much do we know, and how well do we apply what we know? The answer to the first question strikes me as relatively encouraging. Not so the second.

Several important lessons about human resource management are revealed in these chapters. First, they show an overriding concern with the individual, and it is reassuring to presume that this concern pervades the authors' companies as well. Kramer reports that Philips has traditionally been organized around people, rather than the reverse. Morton claims that innovators like working at Hewlett-Packard because "We are an engineering-oriented company. Innovators find stimulating peers, people they can relate to and talk with in their jobs."

Ibuka writes that "Top management strives to create an environment in which each Sony employee—from the highest levels to the lowest—feels rewarded and filled with the joy of working."

If one were looking for a corporate charter designed to produce the kind of environment these comments reflect, one could hardly do better than ponder the "prospectus" Ibuka wrote in 1946 for his new company. Sony's founder wanted a corporation in which:

- All people, particularly technical employees, are respected and are able to work to the best of their ability
- Competitors' products would not be imitated; instead, the goal would be to create consumer products that had never existed in Sony's market, and which would incorporate the most advanced technology.

Sony's charter illustrates two specific lessons which academic research on management has only recently confirmed. One is the importance of specific goals. During the 1950s and 1960s, a number of U.S. managements became infatuated with "the lab in the woods." All you had to do, the gurus preached, was isolate your R&D people in beautiful—but remote—surroundings, provide the best equipment, and avoid pressuring them with deadlines and targets. This done, simply sit back and wait for all kinds of great discoveries to come your way.

Needless to say, it wasn't long before these managements learned that too much isolation and too little pressure stifled rather than stimulated creativity. The labs flopped and were soon replaced by organizations tied more closely to the real world.

The importance of setting goals has since been demonstrated by management experts Edwin A. Locke of the University of Maryland, and Gary Latham of the University of Washington. Twenty years of research involving over 100 experiments in all kinds of work settings have convinced them that goal setting is probably the most effective motivational technique there is.

Ibuka learned this lesson well. "Sony was capable of achieving higher goals, of conquering very difficult technical problems, only if top management gave the targets." His colleague Akio Morita states the case even more strongly in his recent book, *Made In Japan: Akio Morita and Sony* (with Edwin M. Reingold and Mitsuko Shimomura, E. P. Dutton, New York, 1986). Sony's chairman and chief executive officer writes: "Management of an industrial company must be giving targets to the engineers constantly; that may be the most important job management has in dealing with its engineers. If the target is wrong, R&D expenses are wasted, so there is a premium on management being right."

The second concept in Ibuka's prospectus is the importance of assigning engineers to exciting, challenging projects. As Kramer puts it, "Perhaps the strongest motivation can come from an inspiring, challenging project." He therefore urges management to support—and give recognition to—"first class work."

The inspiring, challenging projects at Philips were the megabit memory and the compact disk. At Sony they were the tape recorder, the transistor radio and TV, and the videotape recorder. I daresay Kramer and Ibuka would offer a pessimistic prognosis for any company that couldn't furnish equivalent examples.

A recent study by Thomas J. Allen of the Massachusetts Institute of Technology's Sloan School of Management, and Ralph Katz of Northeastern University, supports Kramer and Ibuka. Allen and Katz found a surprisingly large number of engineers who preferred "the opportunity to engage in those challenging and exciting research activities with which you are most interested" over promotion up either a technical professional ladder or a management ladder. Significantly, this preference grew with age, suggesting there's much that management can do to help older engineers remain creative and productive.

This finding by Allen and Katz calls into question one of the basic assumptions underlying the so-called dual ladder. A number of companies have tried some variation of Kramer's scientific advisor track, but there's been a lot of dissatisfaction with this 30-year-old experiment. Today, it's clear that each organization has to tailor the dual ladder to its own particular situation if it wants to benefit from the concept.

One of the most important insights of the previous chapters concerns the circularity between an environment that is conducive to getting the best out of creative people and the people themselves. Morton and Ibuka point out the extent to which, over time, the environment depends on the people you hire. As Morton observes, "The accumulated impact of bringing top-flight people into the company year after year is hard to beat."

Note that top-flight at Hewlett-Packard implies more than top grades. Although HP wants people from the top 10 or 15 percent of their class, it also wants flexibility, openness, and an ability to work with others. Ibuka insists that a love of work is more important to him than a person's technical knowledge or skill. "Difficult projects give employees the best training and experience; what they have in the beginning does not matter much." At Sony, as at Philips and HP, there's an emphasis on the "ability to harmonize the conflicting interests of colleagues."

The need to find such people makes hiring a tough job. But it's likely to get tougher as a result of the way in which people seem to be changing. A Philips study detects new priorities: "In the past," Kramer writes, "workers were primarily interested in high income and a career... People now give greater importance to developing themselves and realizing their full potential, to being challenged, to feelings of being useful and of belonging to an organization."

In his latest book, *Why Work: Motivating the New Generation* (Simon & Schuster, 1987), Michael Maccoby finds a new type of employee entering the workplace. The newcomer's main goal at work is self-development. Writing in *Research Management* (Jan./Feb. 1987), Maccoby observes that "self-developers are frustrated and turned off by bureaucratic organization and managers who do not share their values...[they] are motivated by managers who give them opportunity, build teams and coach them."

Maccoby, who directs the Project on Technology, Work, and Character, warns that industrial organizations are going to have to change in very basic ways if they are to hire and keep the people they will need to create tomorrow's products and services. "In a work place demanding a combination of technical knowledge and teamwork, these self-developers are motivated to solve problems cooperatively with co-workers, customers, and clients," Maccoby says.

This brings us to my second question about human resource management: How are we doing in applying such lessons? I'm afraid my answer has to be "Not so well."

We have to recognize that the four companies represented in these chapters are by no means typical. They were founded by engineers and innovators, and are still guided by their values. Two of the chapters are by founders. All the founders succeeded in creating cultures that support highly innovative technical work. The companies can boast significant pioneering inventions. They are considered desirable places for engineers to work. HP, for example, is consistently ranked at or near the top of U.S. companies that undergraduates would most like to work for.

But what about the majority of big companies? What about those who still measure their engineers by the acre? What about the 22 leading technology companies surveyed recently by the American Association of Engineering Societies?

The AAES study, *Toward the More Effective Utilization of American Engineers*, found that 90 percent of the respondents placed a high priority on being well utilized; however, only 34 percent felt their companies made good use of their skills. Noting that the findings tracked those of a study he did more than 16 years ago, Fred Landis of the University of Wisconsin at Milwaukee commented that managements don't seem to have learned very much during this time about using engineers.

Particularly disturbing is the prospect that it may become even harder to create and sustain the kind of corporate culture the four authors describe. Noyce notes that the early 1970s were a time of optimism over the economic future in the United States. Today, it's different. ''Our slack, our margin for error, has largely disappeared. Increasingly, we're finding ourselves playing things closer to the vest, not taking the kinds of risks that characterize the big breakthroughs.'' Noyce further warns that unless global economic and political issues are resolved, ''We'll never be able to regain the slack that is so necessary for the individual creative mind to flourish.''

Noyce's words were, of course, written before Fujitsu shook up Silicon Valley executives by announcing its intention to merge its semiconductor business with Schlumberger Ltd.'s Fairchild Semiconductor Corp. (the company Noyce helped start back in the halcyon 1950s). Only a few days later, Hitachi and Fujitsu announced they would team up in trying to break the United States' grip on the world microprocessor market. This reminded industry observers that a recent company history by Fujitsu's chairman, Taiyu Kobayashi, is titled *Fortune Favors the Brave*.

Noyce is right to worry. Consider those companies being forced to ''downsize'' by economic and financial pressures. It's not at all clear that in the long run they will be better off for having had to push thousands of technical (and nontechnical) workers and managers into early retirement. Kramer has already seen the chances for advancement at Philips diminish and believes this will inevitably lead to tensions later on.

It's quite conceivable, therefore, that effective human resource management may pose an even tougher challenge in the years ahead than it did 15 years ago. By suggesting this possibility, and by challenging managers to apply what our best industrial practitioners and academic researchers have learned about managing technical people, Ibuka, Kramer, Morton, and Noyce have done a considerable service.

Part II

A Comparison of Engineering Cultures

The Contributors

Karl H. Beckurts former Executive Vice President and Head of Corporate Technology, Siemens AG, Munich, West Germany
Myron Tribus former Director, Center for Advanced Study, Massachusetts Institute of Technology, Cambridge, Massachusetts; Senior Member, IEEE
Yoshi Tsurumi Professor of International Business, Baruch College, The City University of New York, New York
Michiyuki Uenohara Executive Vice President and Director, NEC Corporation, Tokyo; Fellow, IEEE
George Wise (discussant) Corporate Historian, General Electric Company, Schenectady, New York; Member, The Jovians

6/MANAGING ENGINEERING INNOVATION AT SIEMENS

Karl H. Beckurts

Today we face a situation of rapid technological change reminiscent of the period when electrical engineering and the IEEE were founded. The late nineteenth century was indeed characterized by sweeping technological change. Telecommunication and electrical power reshaped our cultural, social, and political environment.

Microelectronics is making changes today that are no less far-reaching. It has become the basic technology in almost every field of electrical engineering. For two decades, microelectronics has driven progress in data processing, telecommunication, and automation. Under the surface of most industrial products, both new and old, we now find the microprocessor: in cars, household items, telephones, medical instruments, power control equipment, etc.

Because of microelectronics, products are more powerful, less expensive, and less energy-consuming. Extremely high levels of reliability have been achieved, and still higher levels of performance are being developed. New technologies are diffusing more quickly throughout the world, and affecting more and more people.

The European context

Microelectronics has affected engineers more than other people. And it has affected European engineers and their work more than others, for two main reasons.

First, Europe has a long-standing scientific and engineering tradition. For many centuries, science meant metaphysics, medicine, and mathematics. Engineering, however, is different and requires not only knowledge, but also skills. Quite early, appropriate engineering education became necessary in Europe. In Germany, this necessity led to the founding of technical schools. These institutions attained university status—and university quality—through an edict of the King of Prussia in 1899. The oldest of these *Technische Hochschulen*, as they are called, are still operating in Aachen, Berlin, and Hannover. Thus, engineering education in Europe was established in parallel with science education at the universities, but was based on the already existing technical schools.

Despite this formal distinction between "practical science" and "theoretical science," engineering in Europe has always maintained a scientific attitude. This means that engineering quality is not identified just with success in getting a problem solved or a device working. The European engineer's quest is for technical elegance, cost-effectiveness, and reliability in a product—in some cases even for a little too much perfection, perhaps. Achieving this kind of quality takes time, and, to the outsider, European engineering may seem to lack innovative drive. Europeans may seem to be exploiting microelectronics more slowly than others.

The second reason that engineering in Europe is different from that in other industrialized areas is that Europe is not a single body, neither culturally nor with respect to its markets. The cultural diversity of Europe is a valuable asset. But the heterogeneous structure of the European market is a serious drawback.

Microelectronics and other high technologies are flourishing best within large and homogeneous markets. They require major efforts in research and development that often call for industrial cooperation. But industrial cooperation in Europe, as a rule, means international cooperation, in contrast to such cooperation in the United States, which is merely national. Management of international programs is not only difficult, but becomes something of a political venture.

I am deeply convinced that, on the whole, electrical and electronics engineering in Europe is on a level equal to that in other parts of the world. However, the environment in which European engineering is conducted is quite different. Continuity and compatibility are given great importance, and the systems approach has a strong tradition.

Siemens: A leading European manufacturer

Let me be more specific. How is this special character reflected in a leading European electronics company such as Siemens?

The company started in 1847 with 10 people. In 1985, Siemens had nearly 350,000 employees worldwide and sales close to 55 billion Deutsche marks, which makes us the largest German industrial enterprise.

We are not particularly well known in the United States. Nevertheless, we are active in almost all fields of electronics and electrical engineering. Our interests range from microprocessors to power plants, and include heavy electrical equipment, industrial automation, medical electronics, telecommunication equipment, and all types of computers. Except for a short episode in the early twentieth century when Siemens also made automobiles, the company has focused on electrotechnology exclusively.

Currently, Siemens is concentrating on three rapidly growing, highly innovative areas:

- Telecommunication, which has always been one of the company's most successful fields of business. We are active in developing and deploying the Integrated Services Digital Network, ISDN, the internationally standardized digital communication network of the future.
- Office automation, which has much in common with telecommunication but is developing quite differently as a market. This is because in Germany, as in most other European countries, the public telecommunication market is regulated and follows different rules.
- Factory automation, another field in which Siemens has a longstanding tradition.

R&D has always played an important role at Siemens. Werner von Siemens, the founder of the company, worked hard to build a reputation in the academic world for the new discipline of electrical engineering. Laboratories were usually small in his time, but he established a major institute, the *Physikalisch-Technische Bundesanstalt*, which still flourishes in Braunschweig and Berlin, and carries out work like that of the National Bureau of Standards in the United States. The *Bundesanstalt* was probably the first engineering laboratory in Germany. Later, in 1920, the company set up large internal laboratories that achieved international recognition.

Today, 35,000 Siemens people, about one-tenth of the employees, are working in R&D throughout the world. Our R&D spending grows at a rate of ten percent per year, and reached 4.6 billion DM in 1985. This corresponds to about nine percent of our sales for that year and to nine percent of the Federal Republic's total R&D budget as well. Siemens is not only the biggest company in Germany; it is also number one in Europe in R&D. This R&D effort has helped the company achieve remarkable success through innovation in international markets, notably in digital communication, factory automation systems, electromedical equipment, and nuclear power plants.

Engineering at Siemens

Engineering at Siemens reflects the size and diversity of the company and the engineering tradition in Germany. I will illustrate this fact with some examples that highlight differences from the American engineering scene. Incidentally, I will now confine myself to German aspects; the statistics that I will give pertain to the parent company, Siemens AG, not to domestic subsidiaries such as Osram and Kraftwerk Union (KWU), or the Siemens international companies.

Engineering at Siemens is managed predominantly by people with a university education. Through all levels, up to the executive board, about 70 percent of the managers are university trained. Most hold a degree in engineering or the natural sciences. Some—about 15 percent—hold degrees in economics or law. Thus, a university education, even if it does not guarantee a successful career, can be strongly recommended as a means to that end.

A remarkably high portion of employees at Siemens AG—about 15 percent—have an academic education in engineering or the natural sciences, mainly physics. This percentage is double what it was 20 years ago. During the same period, the number of blue-collar workers declined substantially—by about 40 percent at the lowest level of untrained workers.

Naturally, the activity employing the greatest portion of engineers—40 to 45 percent of them—is R&D. Corporate R&D, which is my responsibility, employs about 1,800 engineers; 80 percent of these engineers hold a university diploma and half of these engineers are Ph.D.s. The remaining 20 percent obtained their engineering qualifications at a *Fachhochschule*, a university-type engineering school that has no precise equivalent in the American education system.

By way of comparison, I will give some figures for educational levels for a representative Siemens research site in the United States. We operate research laboratories for telecommunication products in Boca Raton, Florida; we conduct long-term research in information technology in Princeton, New Jersey; and we do research on medical equipment in Des Plaines, Illinois. On the technical staff at Princeton, for example, 80 percent have master's degrees or Ph.D.s; the remaining 20 percent have bachelor's degrees. This distribution is the same as that at our German research laboratories, if one equates those with bachelor's degrees with our diploma engineers from *Fachhochschulen*.

Academically trained engineers at Siemens are found not just in R&D, however. Thirty-three percent of them work in sales and marketing departments. The reason for this large portion is that Siemens sells more than just components and devices; we focus on offering custom solutions to problems, a practice which requires a large amount of technical consulting. For example, more than 70 percent of our sales in 1984 were in custom systems for automation and control, power, telecommunication, and information processing.

Another significant portion—about 17 percent—of our engineers are in manufacturing and quality assurance. This large allocation indicates the importance we attach to developing the highly automated and flexible factory of the future. And I believe that our requirements for highly qualified engineers in this field will increase.

Engineering education in Germany

What in fact is a diploma engineer in Germany? The German education system distinguishes between two types of diploma engineers: those educated at a technical university (*Technische Hochschule*), and those educated at a *Fachhochschule*. The main difference is that the *Fachhochschule* follows a slightly shorter course with a strong orientation toward work in industry. (I understand that the U.S. concept of the community college is roughly analogous.) The technical university, on the other hand, has a stronger commitment to basic science, especially during the first half of the course.

The general idea behind this scheme, originally, was to prepare the mainstream engineer at the *Fachhochschule*, while the technical university educated its students for research and development as well as for high-level management. However, because of the higher social standing of the university diploma, a growing number of students have opted for a university education during the last few years. As a consequence, it is sometimes difficult for industry to find engineers with enough practical experience.

At present, about two-thirds of the 30,000 engineers at Siemens AG were educated at *Fachhochschulen*. But, among newly hired engineers, about half are university educated.

This year, we expect to hire about 4,000 engineers. This is a large part of the total output of the technical universities and *Fachhochschulen* in Germany, especially of electronics and electrical engineers. To help the schools keep pace with the dynamics of current technological innovation and to foster joint projects under real-world conditions, Siemens has substantially increased its cooperation with educational institutions. In microelectronics, for example, industry will need an increasing number of designers over the next few years—designers who are well trained in the use of computer-aided design equipment. Therefore, we have equipped about 20 German universities and *Fachhochschulen* with our CAD system, VENUS, with the understanding that a certain number of student chip designs will be fabricated at our Munich silicon foundry free of cost to the school. The idea is to make the students familiar with the complete development cycle for microelectronic components.

In addition, to meet the company's needs for improved personal knowledge and skills, we have started an extensive program for internal training and continuing education. In 1985, we spent 600 million DM for this purpose. The company runs in-house schools for both technical and management training.

Engineering and culture

Overall, there is a mood of growing optimism in Germany, in contrast to the critical mood of many, especially those in the student generation, in the 1970s.

This does not mean that people are not aware of serious problems. For one thing, we have to reduce the high level of unemployment. We also face the growing menace of environmental pollution. For many people, the speed of technological change has become breathtaking and they cannot understand the reasons behind it. And while we all are hopeful that the world's East-West antagonisms may be calming down, we fear that North-South problems will remain serious.

Nevertheless, a new optimism is in the air. Hope is growing that we will find solutions to these problems—not by halting technological innovation but rather by making wise use of it.

A positive sign is that during recent years the number of students in engineering disciplines has been growing again in Germany. Young people realize that engineering excellence is getting new recognition throughout the industrialized world. They see that it is not only the goals that attracted earlier generations—the largest or smallest, the highest or fastest—that count. Instead, they know that we must aim today for a better fit of technical products and systems into our natural environment to improve the quality of life. With this outlook, contemporary engineers find challenging tasks in the search for new materials, new processes, new ways of recycling.

There is a growing appreciation—not only in philosophical circles but also among engineers—that we live in one world. Interdisciplinary work has permeated our industrial operations. The metaphor of the "two cultures" coined by C. P. Snow was probably a fiction from the very beginning. Today, at least, the idea of interdependence and global responsibility is one of the potent driving forces behind technical and industrial progress.

7/THE QUALITY IMPERATIVE IN THE NEW ERA OF FIERCE INTERNATIONAL COMPETITION

Myron Tribus

The problem with U.S. competitiveness is the quality of U.S. management. As the complexity of products and manufacturing processes increases, the weakness of U.S. management becomes more apparent. For example, data developed by the Boston Consulting Group show that for simple manufacturing processes such as steel or paper production, U.S. manufacturers expend no more labor per unit product than Japanese manufacturers—perhaps slightly less. But for complex processes consisting of hundreds of steps, Japanese manufacturers are far more efficient (Fig. 7-1).

Some people think that the relative inefficiency of U.S. factories is a matter of culture rather than of management. But consider four television plants studied in the United Kingdom by the late Professor Makoto Takimaya ("Japanese Multinationals in Europe: Internal Operations and Their Public Policy Implications," Wissenschaftzentrum, Berlin, Sept. 1979), all employing British blue-collar workers. Two of the plants were managed by the Japanese, one by the Americans, and one—the most highly automated one—by the British. Productivity and quality were best in the Japanese-run factories, and lowest in the British (Table 7-1). The work force shared the same cultural background; only the management was different. Moreover, data now becoming available show that Japanese-managed plants operated in the United States with American labor for Honda, Sony, Nissan, and General Motors achieve the same good results as in Japan. Yes, there is a cultural effect—that of managerial culture.

But management cultures can change. It happened, for example, at Yokogawa Hewlett-Packard, a Japanese partner of HP. In 1975, YHP ranked worst among Hewlett-Packard divisions in profit and defect rate. That year, Ken Sasaoka, president of YHP, and his

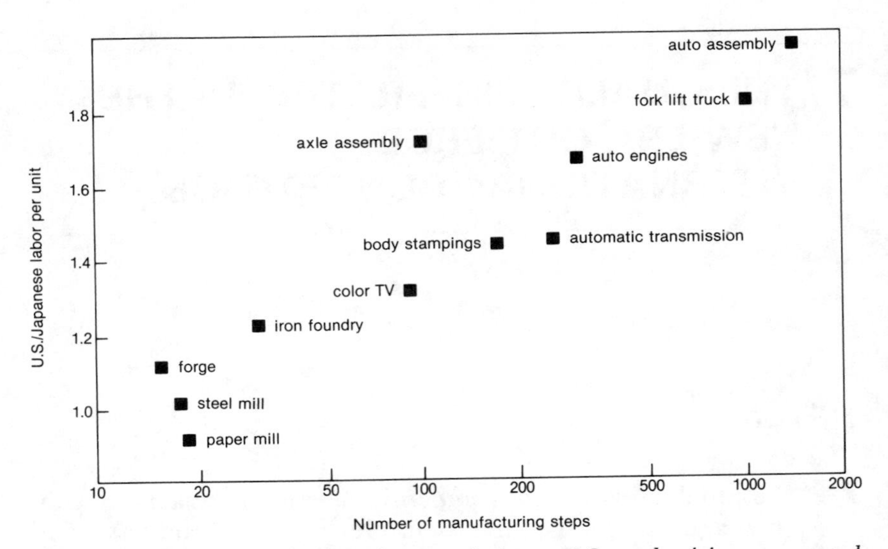

Fig. 7-1: *The effect of manufacturing complexity on U.S. productivity as compared with Japanese productivity is plotted. The vertical axis represents the ratio of labor hours per unit for U.S. manufacturers to that for Japanese manufacturers. (Source: S. Wheelwright, Boston Consulting Group.)*

TABLE 7-1
DATA ON FOUR TELEVISION MANUFACTURERS OPERATING IN THE UNITED KINGDOM

	Management			
	Japanese		U.S.	U.K.
	Plant 1	Plant 2		
Number of employees	700	300	700	2000
Quality index*	5.5	10	13.5	85
Productivity⁺	83	107	71	56
Labor turnover, percent	30	27	30	30

* Percent of sets requiring repair
+ Sets per labor day

(Source: Makoto Takamiya, "Japanese Multinationals in Europe: Internal Operations and Their Public Policy Implications," Wissenschaftzentrum Berlin, Discussion Paper Series, International Institute of Management, September 1979.)

managers decided to do something about it. They studied modern methods of managing for quality and implemented them. As a result of their efforts, YHP changed the division's standing completely. By 1980, the year YHP won the Deming Prize, it had become the most profitable—and most defect-free—division in Hewlett-Packard. (The Deming Prize, named for the U.S. statistician Edwards Deming, who taught the Japanese how to manage for quality in the 1950s, is one of the highest awards for quality performance Japanese firms can win.)

The people who worked for top-running YHP in 1985 were the same ones who worked for trailing YHP in 1975. Sasaoka told me in a long interview that the difference was that he and his people had learned a new way to manage. We followed in fine detail the steps he took and how the employees responded. He convinced me that management culture can be changed—but only if the chief executive officer wants to change and takes the lead in making the change.

The right way to manage should not be called the Japanese way. (As a matter of historical fact, the basic ideas were taught to the Japanese by Americans.) There are plenty of plants in Japan that are managed very badly. When you go to a well-managed plant in Japan or the United States, you find essentially the same situation: the people work together to make things go more smoothly. They study what they are doing, using tools appropriate to the purpose. They are on a path of constant improvement.

Let me give an example of the results of practicing these principles. The Boston Consulting Group has shown that, for a surprisingly large number of products, when unit manufacturing cost is plotted against the logarithm of cumulative production, a straight line results. In other words, when the cumulative production is increased by a factor of 10, the cost goes down by a fixed ratio, for instance 2.

Geoffrey K. Bentley, a student in 1983 at the Center for Advanced Engineering Study at MIT, found an exception to this rule. Bentley examined the operations at Malden Mills' Flock Division and found that cost decreased faster there than the normal BCG learning curve (Fig. 7-2). The reason, he found, was that management and workers worked together to study the process. In so doing they found ways of making the process more efficient.

When you spend time studying what you are doing and use the right tools, you learn faster. That's what the increased slope of the Malden Mills curve represents.

Most of us have heard about the cost advantage that the Japanese have in automobile manufacturing—that they can make a car for about $2,000 less than the U.S. equivalent. Jim Harbour Associates published some data that breaks down the cost advantage into its component parts (Table 7-2). Among the big items are labor,

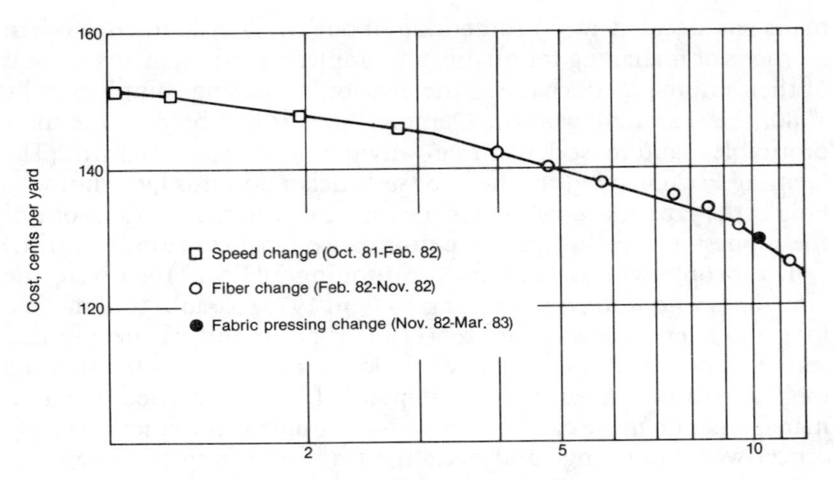

Fig. 7-2: *A fabric manufacturer experienced a faster-than-usual learning rate because of employee cooperation in studying the process and making improvements. The sharp changes in slope represent the introduction of process changes. (Source: Geoffrey K. Bentley, AVCO.)*

TABLE 7-2
COST ADVANTAGE OF JAPANESE AUTO MANUFACTURERS RELATIVE TO U.S. MANUFACTURERS

Advantage	Dollar value
Lower labor time (60 vs. 120 hours)	550
Reduced inspection and rework	329
Lower inventory	550
Fewer job classifications	478
"Shut down line" vs. "tag" operation	98
Advanced technology (robots)	73
Less labor for materials handling	41
Less absenteeism	81
No union representative	12
Total	2212
Transportation	−485
Net cost advantage	1727

(Source: Jim Harbour Associates.)

inspection and repair, and inventory costs—all considerably higher in the United States than in Japan.

I discussed the Jim Harbour data with executives from three auto companies. They agreed that the data, although three years old, are still approximately correct. One of them even suggested that about $1,000 should be added to allow for the undervalued yen.

I cannot quite agree with a statement Lee Iacocca made to a meeting of Tau Beta Pi, the engineering honor society, on October 5, 1985. He said this: "Taking cost out of the car is as important as putting quality in. Right now the Japanese have a cost advantage of between $2,000 and $2,500 per car. And very little of that advantage comes from technology. Most of it comes from the low value of the yen and from the tax advantages their government gives them when they export."

I feel that Mr. Iacocca attributes too much of the advantage to circumstance and not enough to management quality.

The Taylor syndrome

Many people have observed that the techniques used by the Japanese were first developed in the United States, then refined in Japan. The question naturally arises: If the techniques are so good, why aren't they used in the United States? The answer goes back to attitudes formed in the era of Frederick Winslow Taylor, the man who invented time and motion study and is known as the father of scientific management.

Around the turn of the century, Taylor wrote in *Industrial Utopia*: "Hardly a competent workman can be found who does not devote a considerable amount of time to studying just how slowly he can work and still convince his employer that he is going at a good pace." Another quotation from Taylor's book: "Under our system a worker is told just what he has to do and how to do it. Any improvement he makes upon the orders given to him is fatal to his success."

Taylor was a self-educated engineer (as were almost all engineers in his day) who developed methods to increase the cutting rates of metals by proper choice of materials, tools, feeds, and speeds. He turned his attention to increasing industrial productivity using "engineering methods." His first demonstrations, made at Bethlehem Steel, involved teaching workers how to load iron ingots on a railroad car. By teaching them how to lift, when to rest, and how to carry, he doubled productivity. His work established time and motion study. It became the basis for what was to be called "scientific management."

Later in life, Taylor began to doubt his teachings. In fact, he disavowed them in congressional testimony. But his followers had adopted his ideas without question and based their managerial actions on them.

Here is what Mr. Konosake Matsushita, Executive Director, Matsushita Electric Industrial Co., thinks the net result of the Taylor philosophy will be: "We are going to win and the industrial West is going to lose out; there's not much you can do about it because the reasons for your failure are within yourselves. Your firms are built on the Taylor model. Even worse, so are your heads. With your bosses doing the thinking while the workers wield the screwdrivers, you're convinced deep down that this is the right way to run a business. For you, the essence of management is getting the ideas out of the heads of the bosses and into the hands of labor."

Speaking of the Japanese, Matsushita says: "We are beyond the Taylor model. Business, we know, is now so complex and difficult, the survival of firms so hazardous in an environment increasingly unpredictable, competitive, and fraught with danger, that their continued existence depends on the day-to-day mobilization of every ounce of intelligence."

Alternatives

What are the alternatives to the old-style management? Our business schools teach many alternatives, but they are often just fads. I spent 15 minutes in the library to sample some of them and came up with a list of "management by's." Management by:

Agreement
Communication
Compulsion
Exception
Guilt
Information systems
Multiple objectives
Objectives
Obstruction
Participation
Results
Task forces
Walking around.

But I found no mention of management by quality.

Let me describe what management by quality can accomplish. A

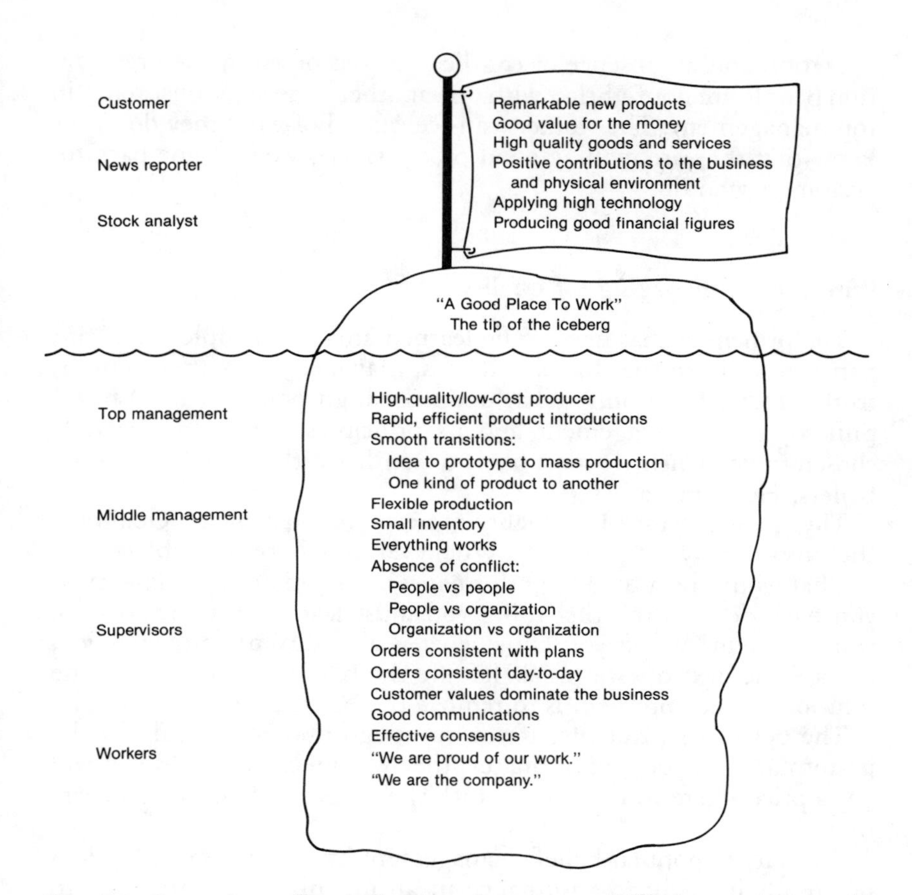

Fig. 7-3: *Perceptions of a well-managed company. Different groups see the company in different ways, but all view it favorably. (Source: Professor Yoshikazu Tsuda, Rikkyo University, Tokyo.)*

well-run enterprise looks different to different people, but looks good to all. To the people outside the company all that can be seen is the results—the tip of the iceberg (Fig. 7-3).

The people outside the enterprise see only the performance: Customers see good products, good value for the money. News reporters see positive contributions and high technology put to work. Stock analysts see good financial figures. But this is only the tip of the iceberg. Beneath the surface there is a structure that supports what can be seen from the outside. Top management's view is of a high quality enterprise—an efficient, low-cost producer, able to introduce new products swiftly and deftly. New ideas are put to work on time and within budget. Middle management sees a smoothly working operation and an absence of conflict; further down in the organization people are not fighting with one another. They see consistency in top management. The workers can feel proud of what they do. They know that they are making good products that won't come back for customers who do.

Principles in managing for quality

The principles that need to be learned are quite simple. The hard part is to unlearn the things that most managers think they know as truths. Here, I can only touch upon the high points of the "new" philosophy of management and its principles. I have deliberately chosen topics which are most likely to conflict with generally accepted beliefs. Here they are.

The quality principle: Quality is never your problem. Quality is the answer to your problem. Whenever you have a problem ask: "What would be a quality performance?" If you do not know, then you have defined the first thing you must learn if you are to solve your problem. When you know what is a quality performance, you can ask the next question: "What are the barriers?" After that, the solution to your problem is to remove the barriers. It is that simple.

The perversity principle: If you set tough numerical goals for the performance of your subordinates, they will meet them and you will pay a price where you least expect it! The system will defeat you every time.

Contrary to popular belief, management by objectives (MBO) does *not* result in higher performance in an organization, compared to what can be done by management *for* objectives (MFO). Re-reading George Odiorne's 1965 book, *Management by Objectives*, in the light of what we know about corporate performance, we see what MBO implies. Whenever you set goals for subordinates to accomplish as a "contract," they are challenged to look good at the expense of the total system. They are encouraged to look good in the short term. More importantly, MBO ignores the redefinition of management. According to the redefinition of management, people work *in* a system. The job of the manager is to work *on* the system to improve it with their help.

MBO submerges concern for system performance under concern for individual performance. No manager has the time and cleverness to compete with people who do *exactly* what is asked.

Most managers do not understand this redefinition. They take the system as they inherit it and try to make it operate. They follow the dictum, "If it ain't broke, don't fix it." They are legitimate prey to competitors who work closely with their employees in a constant quest for improvement.

Most managers do not understand how to go about improving systems. They ask, "What is the best way to control inventory?" instead of asking, "What are the techniques that I can use to improve inventory control systems?" If you accept the revised definition of management, you must learn not only how to improve systems, but also how to lead groups of people in the improvement process. You will need to accept the idea that the people who do the work are the ones who know the most about it. For people who understand this principle, it is not a question of whether worker involvement is a good thing. Worker involvement is a means to an end, not an end in itself.

This redefinition of management's responsibility frightens some people because they are afraid to share power. This is the era of the knowledge worker; sharing power is the only way to get things done efficiently and smoothly in an ever-changing environment.

The process/product principle: If you want to improve the quality of the product, concentrate on improving the quality of the process that produces the product. The quality of the product will then take care of itself.

If you want to improve the performance of an organization, you will need to learn how to identify and improve processes. If you concentrate your attention on the product, you will be looking at lagging indicators. Long before the products become faulty, there are signs in the process that things are going wrong. Fix them and the product will be OK. Most managers have not been taught a systematic way to analyze and improve processes.

Learn to read signals in noise: use statistical methods. Variability of performance is the indicator of a low-quality process.

When a process is not capable of repeating itself accurately, the output will be uncertain and not controllable. All processes exhibit some variation in their performance, if measured closely enough. Managers have to be able to tell when something unusual has happened and distinguish it from variations that should be expected. This ability requires the use and understanding of statistics. Fortunately for most managers, the level of statistical understanding is not much more than required by a good poker player. There are only a few statistical methods, about seven essential ones, each of which can be learned in about one hour. Learning how and when to apply them is another matter.

Managers who do not understand how variability influences the performance of systems, how to measure variability, and how to diminish variability are at an insurmountable disadvantage to those who do.

The principle of continuous improvement: Every activity of people, systems, and machines is a process waiting to be improved. Every process in your enterprise is capable of being improved. Even such a simple task as arranging a meeting is a process. In our office, secretaries have studied this process and developed techniques which ensure that when we hold a meeting, things go smoothly: the cord is long enough for the projector to reach the wall, there is an extra bulb in case one burns out, the people get the agenda on time, and so forth.

Features versus quality

Distinguish between features and quality. Features are what you put into the design of your product to attract the customers you wish to serve. Quality is defined by your customers, and indicates the integrity with which you have provided the features in your product.

Only you can determine the features you will put into your product, for you are in the position of knowing the trade-off between cost and features. On the other hand, only the customer can define quality.

Years ago I purchased a Volkswagen "bug." It did not have a heater, a gas gauge, backup lights, or other amenities. I also owned another car which had seats that moved on command, a radio that tuned itself, an antenna that went up and down, and many fine features. It gave me considerable trouble. My Volkswagen did not have features. My other car did not have quality.

The next person in line is your customer

Keep in mind that the next person in line is your "customer." Teach everyone in the organization that the next person in line is their customer. Teach them to search for ways to permit their customer to do a better job. In many organizations I see that people are intent only on doing their own jobs and, if necessary to meet the MBO criteria imposed on them, will pass on to the next person defective work or do their work in such a way that problems must be solved later on, somewhere else.

This is the quality-cost connection: the producer of highest quality will probably be the producer of lowest cost. Note that I say lowest cost, not lowest price. It is no coincidence that the Honda was the lowest-cost, highest-quality product in its field. When every process by which a product is made is operating efficiently, with everything working right the first time, costs go down, the work proceeds smoothly without error, and productivity is enhanced.

Quality is not an end. Quality is a means to

- Lower cost
- Higher productivity
- Greater reliability
- Less scrap
- Less rework
- Less wasted human effort
- Less wasted physical resources
- Greater customer satisfaction.

Rearrange the process, not the organization

Most people, when asked to think about how work gets done, are apt to first think in terms of an organization chart. But an organization chart is a hierarchical description of how power is distributed in an organization; it does not tell how the work gets done. Friends of mine in the MANS organization in Holland (MANS stands for "Management, A New Style") sent me a diagram that humorously depicts how people tend to describe the workings of an organization. The diagram shows increasingly mindless workers at descending levels of the hierarchy (Fig. 7-4).

Instead of looking at an organization chart, you should study a process flow diagram that depicts simultaneously what work is to be done and how the people enter the process. One of the weaknesses of most modern managerial techniques is that they concentrate either on the people or on the work. Indeed, the "managerial grid" of Douglas MacGregor (*The Human Side of Enterprise*, McGraw-Hill, 1960) postulates that concern for people and concern for the work are orthogonal to one another. But removing the frustrations from the work also caters to the need of the people to get satisfaction from their employment. In the new approach to management, these concerns are not orthogonal.

In a flow diagram of an aspect of the purchasing process (Fig. 7-5) it is easy to identify possibilities for improving the process. The top of the diagram provides a "people coordinate." An activity positioned below a particular person or organization indicates responsibility for the activity. A horizontal line depicts a "customer-supplier" relationship. A diamond (for a decision) indicates a step in which something can go wrong if the person asking for the decision

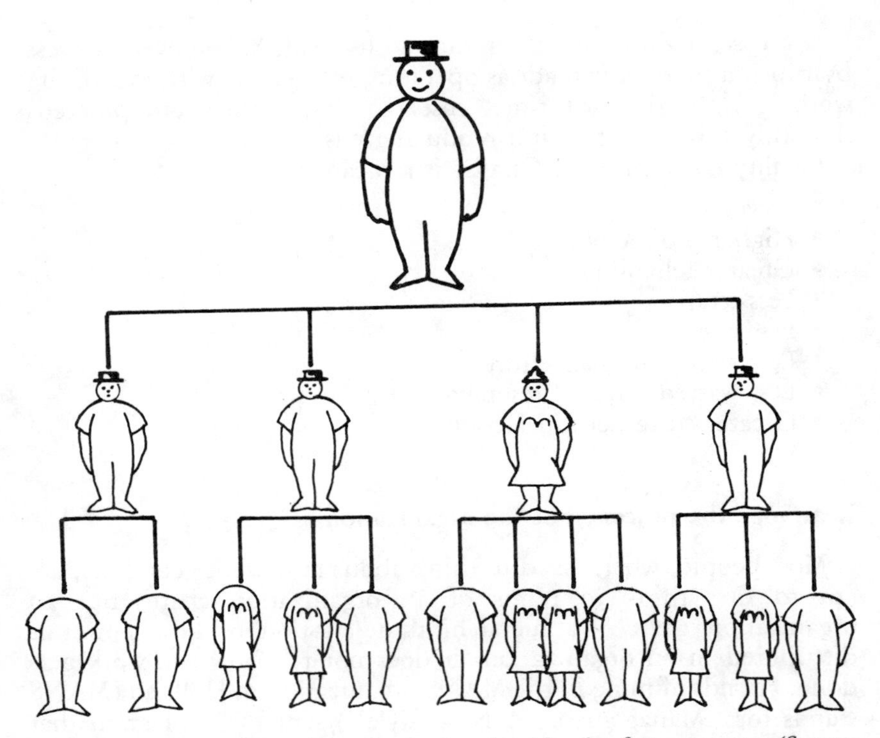

Fig. 7-4: *An organization chart representing the old style of management. (Source: MANS Organization.)*

does not understand the kind of information the decision-maker needs. For example, when a request for purchase is sent to a possible vendor, it is helpful to realize that the vendor is the customer for the RFP. The person preparing the RFP should determine what is the right information for the vendor so that a responsive and useful proposal will be forthcoming.

It makes a difference how you approach the task of making things better in your organization. If you approach it by thinking in terms of an organization chart you will think and act differently than if you approach it by studying the process. If you concentrate on the organization you will

- Try to motivate people
- Ask ''Who is wrong?''
- Define responsibilities
- Watch the bottom line
- Measure people

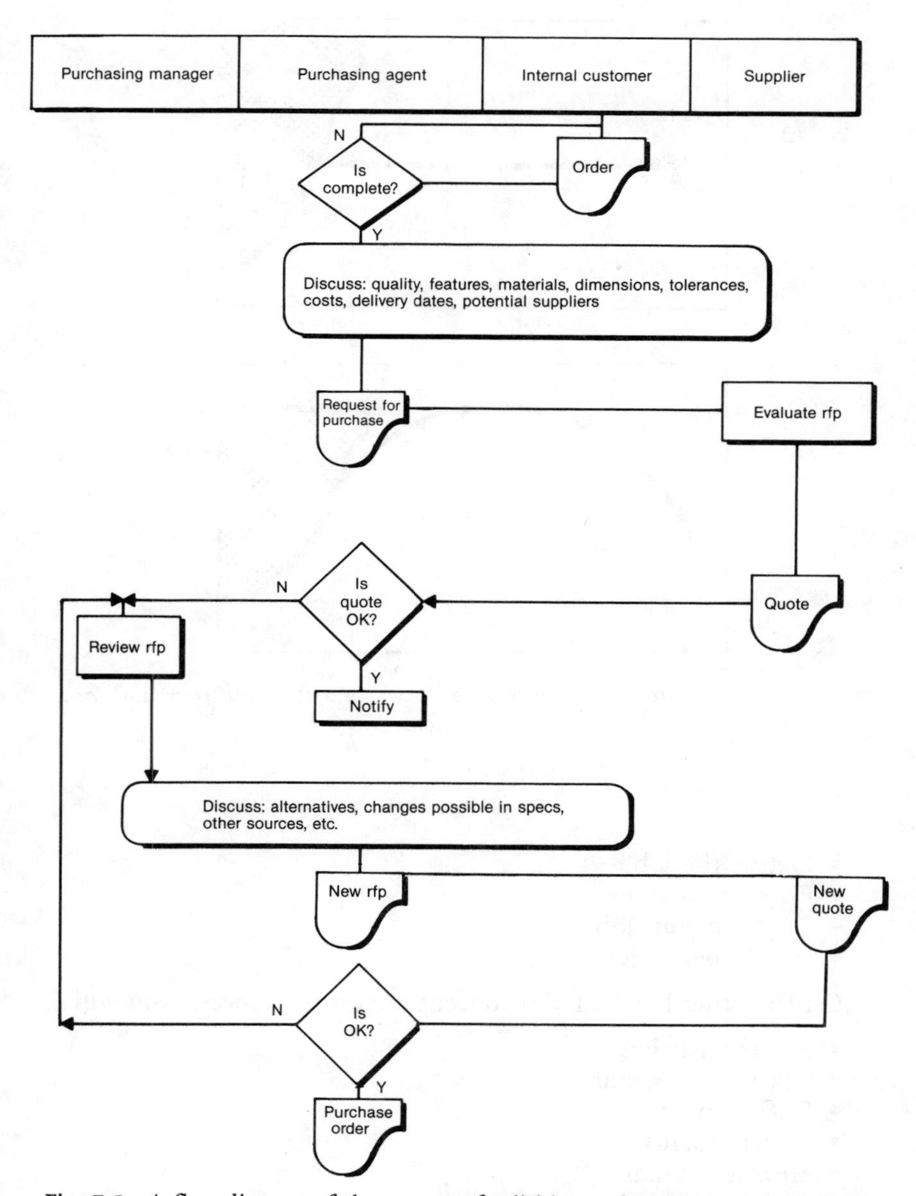

Fig. 7-5: *A flow diagram of the process of soliciting and evaluating a request for purchase. A horizontal line indicates a supplier-customer relationship between the persons or organizations in the blocks at the top.*

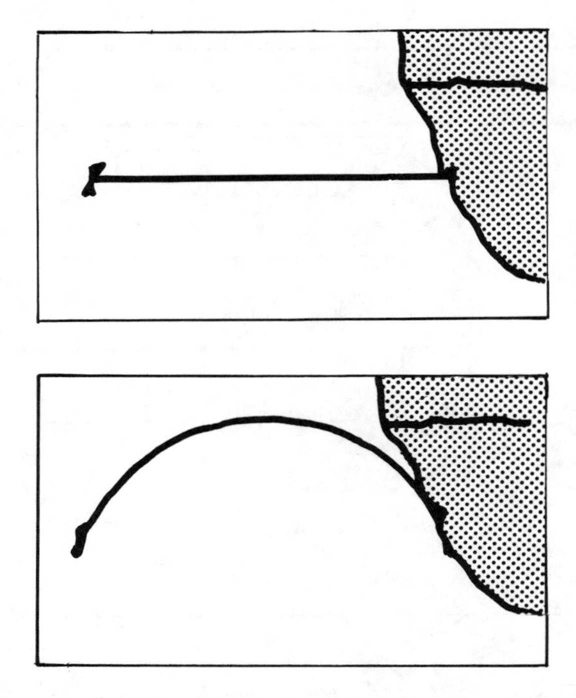

Fig. 7-6: *The route from San Francisco to Tokyo for a flat earth (top) and for a curved earth (bottom) on a Mercator projection.*

- Define the job
- Fix all deviations
- Say ''Do your job''
- Say ''Obey orders.''

On the other hand, if you concentrate on the process, you will

- Remove barriers
- Ask what is wrong
- Define quality
- Watch quality
- Measure systems
- Define the customer
- Reduce variability
- Ask ''Can I help you?''
- Say ''Help me improve things.''

A concluding fable

Once upon a time there was a captain of a ship that carried cargo from San Francisco to Tokyo. He set his compass on Tokyo and stayed on course all the way. One day a passenger by the name of Deming came aboard and suggested that the captain follow a different route (Fig. 7-6). The captain was not interested. ''Everyone knows the shortest distance between two points. I do not have the time or the fuel to go in a curved line. My customers will not wait.''

Deming got off the boat in Tokyo and taught the Japanese captains how to navigate. After a while the captain noticed that his Japanese competitors were advertising lower costs and faster service from San Francisco. When he was in Tokyo he demanded to inspect the Japanese vessels. He found they were of the same design as his own. He did notice however, that the crew was happier (they were in port more often). ''That's it,'' the captain exclaimed, ''it's cultural!'' The captain examined everything except the conceptual framework in the Japanese captain's head.

If you think the earth is flat, you will navigate in a certain way and no one will be able to explain navigation to you differently.

If you want to learn how to manage effectively in the new era, you will need to reconsider your entire conceptual framework. What is at issue is a conceptual revolution as great as when humankind changed its view of the earth and thought of it as a sphere instead of a plane. It took a long time. As W. Edwards Deming has observed, ''It doesn't matter when you start, as long as you begin at once.''

8/THE BEST FEATURES OF EUROPEAN, JAPANESE, AND UNITED STATES CORPORATE AND ENGINEERING CULTURES: ARE THEY TRANSFERABLE?

Yoshi Tsurumi

Xerox's comeback and corporate cultural revolution

Viewed in perspective, Japanese competition in the U.S. has been a blessing in disguise for some enlightened firms like Xerox. Xerox's story began like this. Around 1980, in the high-speed copier market, IBM was squeezing Xerox out, and in the compact copier line Japanese firms were overwhelming Xerox. After decades of operating like a monopoly, Xerox had become too internally rigid to compete in the new, fast-moving technological environment. So Xerox underwent the corporate equivalent of a cultural revolution from top to bottom. This makeover included a major dismantling of departmental turfs and had the end result of motivating Xerox employees to communicate and work together across different functions, from R&D to manufacturing, marketing and vice versa. Xerox had learned many lessons from its joint venture in Japan, Fuji-Xerox. Xerox cut its dealers in the U.S. from 5,000 to 300 and concentrated on providing prompt customer service. By 1985, Xerox's slide in the U.S. market share had stopped. Today, the firm is better equipped to take on IBM and any Japanese competitors. Meanwhile, Xerox has cut its component rejection rates from a whopping 8,000 parts per million in the early 1980s, to 1,300 parts per million by 1985. It has still not matched its Japanese competitors' record of 1,000 parts per million, but it is getting there. Now, Ford Motor Co. is trying to apply the lessons Xerox learned to its firm.

The Xerox saga illustrates what it takes to become competitive in a high tech information age. Without fundamental changes in corporate and engineering culture, no firm can remain competitive and survive today. Firms must learn to adapt to an environment in which the line between domestic and foreign markets is rapidly disappearing. This is the surest sign that the fast moving high tech information age has arrived.

Adaptive corporate behavior that is needed for a high tech age must be based on technology-driven, rather than the finance-driven, strategies. Since technology can only be embodied by human beings, technology-driven management must always be human resource-driven. Rather than chasing money, both management and engineers must cultivate the mindset to chase products and markets. If you chase money instead, you will soon lose it.

American "neutron bomb" executives and the Japanese Paradox

What is hidden in the U.S.-Japan trade dispute is that many American executives wonder if American firms will be able to compete at home, much less abroad, under the new market conditions of the high tech age. In particular, the American management obsession with bottom-line figures has already proved fatal. Nowhere is this problem and its solution more apparent than right here in the United States in the Japanese Paradox. At the first sign of intensifying competition even in the growing U.S. market, just like a real neutron bomb, many American executives zap their people to salvage bricks, mortar, and machines and leave the United States. By contrast, their Japanese competitors are coming to the United States in droves to produce quality products with high wage American labor. This is the Japanese Paradox. As shown by Japan's Bridgestone Tire Company's takeover of the former Firestone plant in Tennessee, Matsushita's acquisition of Motorola's Quasar in Chicago, and Hitachi Metal's purchase of the former General Electric plant in Michigan, some Japanese firms acquired the same plants that their American competitors abandoned as uneconomical and have turned them around successfully. The Japanese Paradox is not black magic or economic judo.

As U.S. market demands continue to shift dynamically, many American firms' overseas plants are isolated from these market changes and are behind their Japanese competitors, who are quickly filling U.S. market demands from their U.S. bases as well. This paradox has also been shown by the Michelin plant in South Carolina.

The Japanese Paradox, Michelin's story, and the recent revival of German and Dutch electronics firms, all show one important ingredient in the successful adaptive corporate system congruent to the high tech age. "Non-neutron bomb" executives of leading

Japanese, German, Dutch, and French firms seek technical and marketing solutions rather than financial, legal, and political solutions. Above all, these non-neutron bomb executives hone their manufacturing skill and fuse their renewed flexible manufacturing skill with their marketing efforts and vice versa. This is the lesson that Xerox has successfully taken to heart.

Adaptive corporations and technology-driven management

At the time of such an economic crisis as a temporary sales decline, neutron bomb executives fire 40 $20,000/year employees in order to preserve their own $800,000/year salaries and perks. They treat human beings as disposable costs rather than renewable assets. As one firm after another fires its managers, engineers, and ordinary employees, misery spreads at compounded rates throughout society. Unemployment becomes entrenched. The trade deficits and federal budget deficits multiply. The social and economic inequities worsen.

Unfortunately, during the past 40 years, many American and European corporations have forgotten what made American and European economies strong and internationally competitive. In particular, from the turn of the century to the 1950s, the United States became the world's unquestioned industrial leader by developing what I call the triad of the technology-driven economy. That is to say, the United States was blessed with the simultaneous and mutually reinforcing developments needed for modern technology, namely, (1) product-specific innovations, (2) production process-specific innovations, and (3) the management and entrepreneurial commitment to combine product and production process innovations with human beings. Unfortunately, the immediate postwar supremacy of U.S. industries lulled American management and engineering thinking and training to such overconfidence and passivity that it destroyed the triad of the technology-driven economy.

The legal, financial, and bureaucratic manipulation of the corporation and its employees came to be equated with corporate management. Consequently, unlike the previous era, from the 1950s to the 1980s over two-thirds of the chief executive officers (CEOs) of the Fortune 500 blue chip corporations in the United States have risen from the ranks of those with expertise limited to financial, legal, and advertising deals. Meanwhile, one engineering school after another has tried to suppress the hands-on tradition of real world problem solving of the engineering profession. Instead, pure scientists or, at worst, "design engineers" have become the role models. In the

United States, therefore, no self-respecting engineer would be caught dead mingling with technicians and production workers on factory floors. Many American engineers have come to believe that their task is to design products and production processes without ever considering how their new products will be made at factories and how their products will be greeted by customers. They do not wish to be associated with such lowly creatures as sales engineers and industrial engineers, much less with field sales forces. Thus, a firm's R&D and manufacturing activities have come to be separated completely from its marketing activities and vice versa. As a result, many R&D activities have come to emphasize product innovations for innovation's sake without much prior thought given to their manufacturing and marketing applications.

Meanwhile in Japan, from the late 1940s to the 1950s, more and more firms were embracing the lessons learned from three American engineers and one American statistician. This statistician is the famous Dr. Edwards Deming. But preceding Dr. Deming in Japan were three American electrical and electronics engineers, Frank Polkinghorn, Charles Protzman, and Homer Sarasohn, who were with the Civil Communication Section of General MacArthur's headquarters. These engineers helped Japanese executives and engineers to understand the management leadership role committed to their industrial growth through the quality–first approach. Their Japanese disciples took the lessons to heart and went on to develop what has become by now the renowned "Japanese total quality control system" under the tutelage of Dr. Deming. Out of this total quality control system emerged the Japanese flexible manufacturing system. It should be noted that Polkinghorn, Protzman, Sarasohn and their Japanese counterparts were bona fide electrical and electronics engineers, not industrial engineers. But they led Japan's head-on commitment to improving its competitiveness in worldwide markets through productivity and quality improvement. Again, in 1950, when the first of Dr. Deming's seminars on statistical quality control was given, all the participants were engineers and executives. But their first hands-on lesson was to conduct door-to-door consumer surveys to investigate Japanese housewives' complaints about Japanese sewing machines. In short, the market problem orientation was made the foundation of Japan's quality control drives.

Under Japan's new corporate system, over two-thirds of the blue chip firms have come to be headed by individuals with science, engineering or actual sales expertise. Furthermore, according to one survey by a British firm in 1985, today 28 percent of Japanese firms have at least one corporate board member designated for scientific and technological forecasting and assessment. By contrast, only six percent of American and three percent of British counterparts respectively have such board members. In Germany and France, where there has always been a strong traditional respect for scientists and engineers, the board makeup is closer to the Japanese situation.

The modern Japanese corporate system, which is definitely of the postwar vintage, invariably requires that in times of economic difficulty, top management cut its salaries and perks first before rank-and-file employees are asked to make temporary salary concessions. This method of prompt adjustment and adaptation through variable income sacrifice from top to bottom is the foundation of Japanese firms' adaptability to the volatile economic and technological environment worldwide. This is in stark contrast to the expected behavior of American management.

Since the high tech information age requires firms to make continuous adaptations, made possible only if workers acquire newer and newer technological and management skills and views, it is no wonder that neutron bomb executives soon destroy their firms's competitiveness at home and abroad.

The fallacy of the bottom-line obsession

Let me explain how Japanese firms's rejection of the obsession with bottom-line figures led to the zero inventory system, or flexible manufacturing system, and yielded greater profits as well. According to the prevalent theory of "bottom-line maximization," the fewer inventories you carry, the better. Many operations research formulas such as the *economic order quantity* (EOQ) formulas have been developed to help firms find the "optimum" inventory level.

All the EOQ formulas I have studied are based on one fallacious assumption: even if you lose sales because of insufficient inventory, the same customer will show an equal probability in the future of returning to you, not to your competitors. As you know, of course, in the real world of branded high tech products, once you lose your potential customers to your competitors, you rarely get them back. For example, in the United States today, if you lose an auto or computer sale because you do not have the product that a potential customer wants, it will be seven or eight years before that particular customer will be in the market again. Meanwhile, you lose all the interim streams of product service revenues and profits. If that customer is even mildly satisfied with your competitor's product, he or she will most likely stay with your competitor. Worse yet, his or her friends will also be drawn to your competitor through his or her word-of-mouth advertisement of your competitor's product. Accordingly, market competition in the real world should tell you that you can never have enough inventory. The more inventory you have the better.

But it is too costly to carry larger and larger varied inventories. How do you solve this dilemma? The solution is not found by revising EOQ formulas or rewriting computer simulation models of inventory and production controls. The only real solution is found in management's commitment to finding simultaneous answers to your needs of satisfying your customers at all times while cutting inventory costs and risks of product obsolescence. The final and technical solution is a zero inventory system.

Once you recognize this simple truth, you have to build flexibility into your own production systems to make just-in-time deliveries of your products to your dealers and customers. You have to make these products right the first time and all the time. The flexibility of your manufacturing systems, not the inventories, will cushion the uncertainties of market demands. Since you have no inventories to cushion dynamic market demand shifts, you would notice market demand shifts immediately. Since you do not have work-in-process inventories to hide your imbalances of production lines, you would notice immediately the production bottlenecks and move to correct them. This way, technological innovations are also made market-driven.

It is true that many more Japanese firms than their American and European counterparts possess adaptive behaviors with the necessary prototypes of management and engineering culture. However, this does not mean that Japan will win the high tech race for industrial supremacy. Japan now excels in mass producing quality products flexibly enough to keep up with dynamic market demands worldwide. But Japan now faces the new challenge of catching up with the United States in the innovative software designs of computer-related technology and other high tech products.

Some American observers draw undue comfort from the fact that the Japanese are struggling with microprocessor designs and artificial intelligence development, for example. Some say that the Japanese are good at mass producing quality 256K RAM chips, but never at software designs of microprocessors and other high tech products. Little do they know, however, that the Japanese have been underestimated before in their ability to upgrade skills in other fields such as steel, automobile, ship, consumer electronics, 16K RAM chips, 64K RAM chips and hosts of manufactured goods and industrial innovations, only to be disproven.

The Japanese are keenly aware of their weakness in software designs of computer-related technologies and other information age products and their uses. Therefore, just as they did in the past for steelmaking, shipbuilding, and consumer electronics, they are pooling their human resources from government, industry, and academia to tackle these problems. As we all know, the solution of the problem begins with an awareness of it. Once the Japanese society is awakened to the next hurdle of its technological challenge, it will demonstrate a single-minded pursuit of its solution. In order to accomplish such a goal, the Japanese have in the past changed drastically their educational and corporate systems. Japan has come to accept the fact that the industrial structure is always in a state of flux. Even in order to stand still, society must keep on moving. The United States cannot afford to be complacent about its lead over Japan in the product-specific technologies of microprocessors and other high tech products.

International transfer of adaptive corporate systems

All the examples that I have shared with you have one thing in common: in the high tech information age, the adaptive corporate and engineering culture is universally applicable in any country situation. During the last 40 years, Japanese firms have embellished what they learned from the American industrial experience for the first half of the 20th century. Meanwhile, some German, Dutch, and French firms have also thrived with their own triad of technology-driven growth in world markets. Despite the spreading neutron bomb behavior in the United States, there are over 40 firms in various industries that have retained the best tradition of adaptive corporate behavior. Furthermore, the Japanese Paradox in the United States is proof that Japanese manufacturing skills and their accompanying corporate and engineering culture are sufficiently transferable to the United States. The same observations can be made of Japanese subsidiaries in the United Kingdom and continental Europe.

Accordingly, what we must ask is not whether the so-called "Japanese," "European" or "American" corporate systems are internationally transferable, but how best we can change the current thinking of firms mired in the neutron bomb mentality. What are appropriate management and engineering viewpoints and the resultant corporate system of adaptive and competitive behavior?

To begin with, quality control, flexible manufacturing systems, and marketing should not be treated like manipulative tools. Certainly, there are technical tools and skills essential for success in quality control, flexible manufacturing systems, and competitive marketing. However, what matters is the corporate and engineering cultural context in which these tools are applied by managers, engineers, and employees to their daily tasks.

Secondly, and most importantly, adaptive corporate behavior requires true entrepreneurial leadership. Self-sacrifice is a necessary leadership quality. Although American business and engineering schools do not teach this, courage, high purpose, honor, and independence form the basic values of entrepreneurial leadership. In 1950, Charles Protzman taught Japanese executives and engineers that a leader's main obligation is to secure the faith and respect of those workers under him. "A leader could gain," said Protzman, "his workers' support simply by being loyal to them." At present, the Toyota-GM joint venture in Fremont, California, has pledged that management will cut its salaries before rank-and-file employees are asked to make temporary economic concessions.

Thirdly, adaptive corporate behavior requires a constant process of education for everybody from the president down to the floor sweeper in the factory. The education content must therefore consist of three intertwined elements: (1) updating whatever skills and knowledge each person needs to perform his or her job of adaptive corporate system, (2) at the same time, expanding each person's ability to work with other human beings, and (3) comprehending how delicately the United States has become politically, technologically, and economically interdependent with the rest of the world in general and with Japan and other Pacific nations in particular. These three ingredients must form the core curriculum for the continuous training of engineers, managers, and rank-and-file employees.

Currently, there is no engineering or business school in the United States that teaches the above three vital components of the continuous education of any professionals. The remedy will therefore be found in the special model training curricula of a professional group like the IEEE.

References

[1] Y. Tsurumi, *Multinational Management,* 2nd Edition. Cambridge, MA: Ballinger, 1984, chs. 1, 4, 11, 13.

[2] Y. Tsurumi and H. Tsurumi, "Value-added Maximizing Behavior of Japanese Firms and Roles of Corporate Finance," *Columbia Journal of World Business,* Spring, 1985.

9/UTILIZATION OF ENGINEERING PERSONNEL

Michiyuki Uenohara

The key to competing successfully in the electronics industry is overall strength—each link in the chain from basic research to development to production to marketing must be as strong as the next. I believe that overall strength is the secret of the success of Japanese industry in general and of the NEC Corporation in particular. Western Europe may do better research, and the United States may do better marketing, but a chain is only as strong as its weakest link. Japanese industry makes sure that there are no weak links and no exaggeratedly strong ones—that would be inefficient.

How does Japanese industry do it? How do we forge links of uniform strength? Primarily, we do it by effective utilization of our engineers: by distributing highly capable engineers throughout the company, by assigning them personal responsibility, and by making sure that everyone—researcher as well as sales engineer—is sensitive to the market.

Before I elaborate on this strategy, let me give some background on my company. NEC was established in 1899 as a joint venture between AT&T's Western Electric Company and Japanese partners. (Western Electric's ownership was transferred to ITT in 1925 as a result of an antitrust action against AT&T by the U.S. government.) NEC's current major businesses are communications, computers and industrial systems, electron devices, and home electronics. These areas contributed, respectively, 29, 32, 27, and 8 percent of our sales in fiscal 1984.

That year, 34 percent of our sales were exports to more than 140 countries. Sales totaled $7.5 billion for the parent company, $9 billion if subsidiaries are included.

The parent company handles administration, R&D, engineering, and marketing and sales. The subsidiaries are responsible for mass production of established products. We have fully owned subsidiaries in Japan and 12 other countries, including a plant for very large scale integrated circuits in Roseville, California.

About 11,500 engineers work for the parent company and another 10,500 in the subsidiary companies. Of the total, about 12,000 are hardware-oriented engineers and about 10,000 are software-oriented.

These engineers are the most important resource at NEC. Utilizing them effectively is crucial to carrying out our corporate strategy for maintaining competitiveness. Out policy toward them, as it is toward all employees, is embodied in three principles: making full use of individual ability, encouraging flexibility, and ensuring fair treatment. We are careful to maintain high levels of motivation among our engineers through compensation schemes at least as good as those of local competitors. And we help them to develop career paths that suit their interests and abilities.

NEC is a leading employer of engineers in Japan. Since 1981, new engineering graduates have ranked NEC as the company they would most like to work for.

The NEC tree

To emphasize to engineers their key role, we describe our company to them in terms of a tree (Fig. 9-1). The roots represent technology, and the branches are our four business areas. Above the tree is the sun, symbolizing our customers. As the roots gather nourishment and send it coursing through the trunk, the branches interact with customers. Growth ensues, opening up new fields in every walk of life—in government, in business, in the home. Engineers can aid growth by communicating and cooperating with each other in a timely manner for customer satisfaction.

To encourage communication and cooperation, we have distributed NEC's R&D activity throughout the company. The Research and Development Group and the Production Engineering Development Group, which together constitute the corporate R&D laboratories and which I am responsible for, belong to the operating organization, not the staff organization (Fig. 9-2). The six profit-making operating groups have their own divisional laboratories and development departments.

The Research and Development Group is primarily engaged in long-term research in technologies for products—not tomorrow's products, but the day afte: tomorrow's. Every manufacturing division is responsible for developing marketable products—today's products—as well as the techniques necessary for productivity and quality. And the divisional laboratories work primarily on technologies that

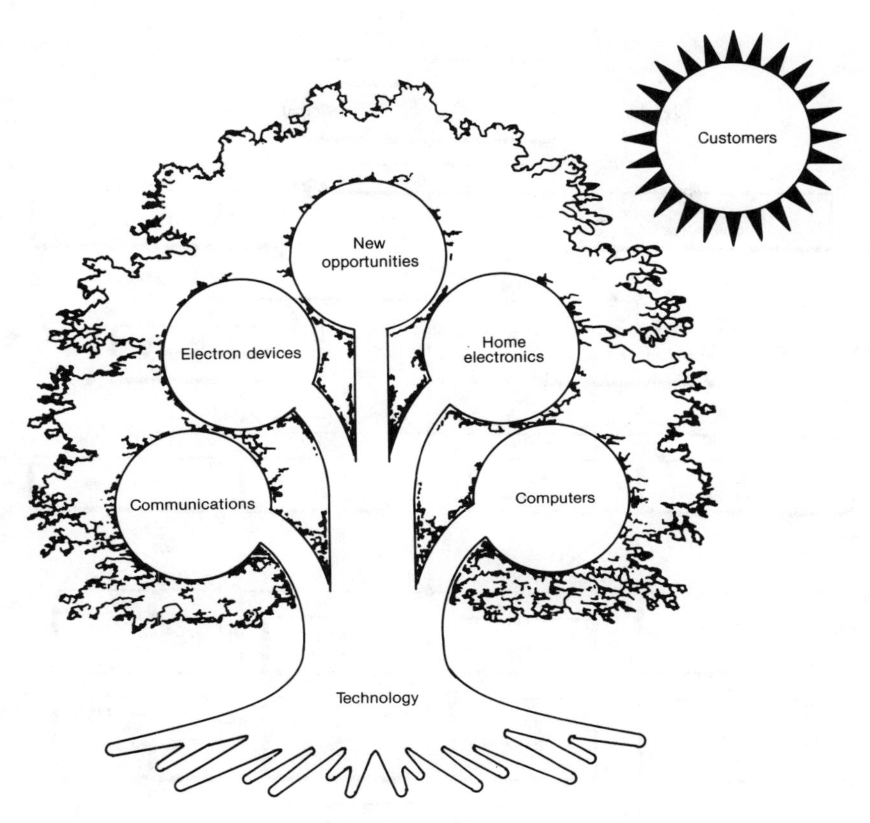

Fig. 9-1: *NEC tree.*

will create the group's near-term business—technologies for tomorrow's products. Of course, product technology and production technology are inseparable; even research for the day after to tomorrow must proceed amid a strong awareness of production realities.

NEC employs some 36,000 people, of whom about 32 percent are engineers. More than 8,000 NEC engineers participate in research and development. About 1,000 are in the corporate laboratories, about 2,000 in the divisional laboratories, and about 5,000 in the manufacturing divisions. Most of the remaining 3,500 engineers in the parent company are involved directly in manufacturing; a few are in the sales groups. At NEC, however, since the manufacturing divisions are the profit centers, most engineers have to participate in marketing, especially for new products and systems. Even researchers often accompany division engineers or salesmen to support their

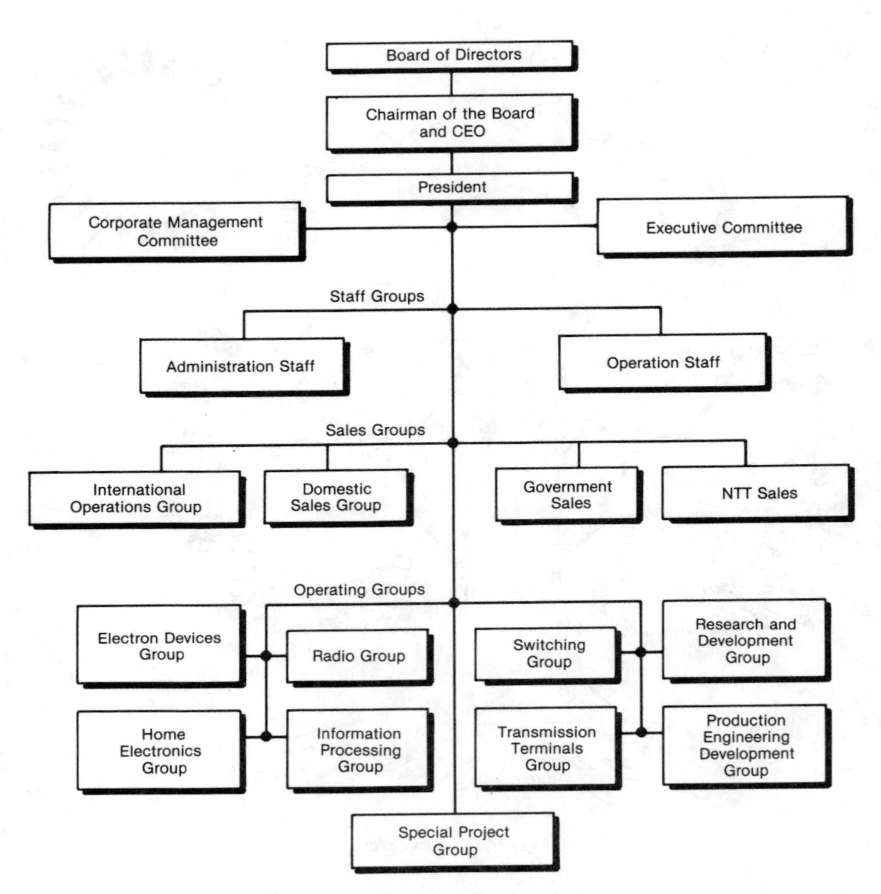

Fig. 9-2: *NEC organizational chart.*

marketing activities. In subsidiary companies, about 3,000 engineers participate in R&D.

Allocating for today, tomorrow, and the day after

Of course, it is difficult to define boundaries between technologies for today, for tomorrow, and for the day after tomorrow. Even if they could be defined readily, the boundaries would be continually changing as the market and the technological environment change.

To cope with such a complex situation, NEC has adopted a policy of selective distribution and concentration of engineering resources. It seeks to balance the demands for R&D in a way that promotes the continued health and growth of the business.

Consider the needs for education of engineers and for research to establish basic technologies—both require time-consuming effort and long-range planning. On the other hand, development of new products and of methods of producing them must be done in a responsive and timely way to meet market demands. The latter activities consume an even larger share of engineering time and money. Indeed, if a thoughtful strategy is lacking, the education and basic research would be put aside and the commercialization activities would consume an even larger share of resources and time. Our distribution and concentration strategy aims to avoid such chaos.

Setting the strategy for R&D programs involves high-level interaction among marketing groups, operating groups, and the Research and Development Group. First, all new product ideas are analyzed to determine their future market potential—analyzed in the light of long-range corporate objectives. For the ideas that appear attractive on the basis of market criteria, the company decides what technologies would be required to achieve the products and what core technologies should be developed in-house. The decisions on core technologies, together with strategic technology domains, are presented to all operating groups, and the selected technologies are committed to nurturing for 10 or even 20 years. Emerging new technologies are pursued by the corporate laboratory and established technologies by the manufacturing groups.

Since basic technologies that constitute a core technology change more frequently, the selection of basic technologies and R&D approaches are subjected to initiatives by each technology group. Initiatives by individual engineers are encouraged. This distributed activity on basic technologies is coordinated through cross-organizational communication and cooperation.

Distribution and concentration

Later, when the time comes to develop new products on the basis of new core technologies—or when basic technologies can aid in a competitive crisis—such technologies and specialists from several organizations are concentrated where they are most needed. This is the "distribution and concentration" concept of managing engineering resources. It assures that objectives will be accomplished in the shortest possible time, since it ensures that engineers become well-versed in a given technology and its ramifications throughout the company, then it brings their skills and expertise to bear on the real-world problems of converting the technology into a product.

However, for the distribution and concentration strategy to

function successfully, interorganization communication and cooperation is essential. To make it easier for engineers in different organizations to talk and work together, we have taken the step of completely eliminating hierarchical relationships within and among different groups. All engineers are treated equally, regardless of classification. We find that this works well; when engineers respect each other for knowledge and professionalism rather than for titles, the project advances more rapidly.

Investing in engineers

Even today in Japan, engineers tend to stay with an employer for a lifetime. Thus, when NEC hires a new engineer, the company is committing itself to an investment of around $2.5 million in salary alone.

Besides this financial commitment, the company has a responsibility for continuing the education of engineers in order to extend their professional lives as long as possible. Toward this end, NEC has extensive programs. On-the-job training and self-education are probably the most effective means, but they tend to limit an engineer to his own specialty. To broaden the engineer's scope and strengthen basic knowledge, the company maintains the Institute of Technology Education for engineers, the NEC Technical College for technicians, and the Institute of International Studies for managers and other employees who travel abroad.

Engineers are also encouraged to attend conferences, meetings, and seminars. Several thousand engineers travel abroad every year to attend international conferences and committee meetings and to visit foreign customers and partners.

NEC and innovation

Effective utilization of engineers will grow ever more important for NEC in the years ahead as the company relies more and more on newly developed technology. In the past, Japanese industry has been criticized as an imitator. There was some truth in this, and there was good reason for Japan's noninnovative approach. For many years, Japan sought primarily to satisfy domestic needs, and the most resource-effective way to do it was to adapt and improve existing Western technology. But now basic domestic needs are beginning to approach saturation, and new needs are beginning to arise—those of a worldwide information-oriented society. Technological innovation will be the key, and we must continuously motivate our engineers to accept the challenge of innovation.

But we won't lose sight of our roots; we are not going to alter our need-oriented approach. Technology, like Mount Fuji, is beautiful not for its heights alone but also for the extended foothills of hard-won improvements in quality and cost of popular products.

10/MANAGING IN A COMPETITIVE ERA

George Wise

In the 1980s, the question "how should you manage innovation?" has taken the form "how do you match Japan?" Answering the question is now passing from a first phase of imitation and panic to a second phase: calls for radical reform. The previous chapters are evidence of that transition, indicating the form those reform proposals will take, and even reminding us that the phase of radical reform is likely to give way to phase three: reform meets reaction.

That first phase reached full swing when Japanese success became too pronounced to be attributed simply to lower labor costs; that is, when victories in steel and automobiles were followed by triumphs in high tech territory: first consumer electronics, and then semiconductors. Each advocate of imitation picked a different Japanese institution to which he attributed that nation's success, and advocated its immediate adoption in the United States. It might be something limited and direct, such as quality circles, or just-in-time inventory systems, or something global and indirect, such as the role of Japan's Ministry of International Trade and Industry (MITI), or Japanese investment banking practices.

The purveyors of panic, on the other side, threw up their hands and proclaimed the uselessness of trying to match the Japanese miracle. It was based on that mysterious element, national culture. The Japanese worker had been programmed by culture to do wondrous things that Americans could not or would not match. Any evidence of Japanese cultural difference, no matter how absurd, unlikely, or of dubious social value, was pointed to as evidence of Japanese cultural superiority. Thus, one commentator pointed admiringly to the alleged fact that on any given school day in Japan, every student in each grade is studying exactly the same page in exactly the same textbook. Another recalled how, during the visit of an American management delegation to a Japanese factory, the workers willingly gave up their morning break so they could be seen hard at work at their machines when the American visitors trooped through. "Imagine," he concluded, "if we'd asked our workers in" (here he named two of his company's rust bowl factory cities with strong unions) "to do that!"

Such anecdotes become alibis: sure, we managers could have beaten Japan—if only our students and workers had cooperated! (These alibis are reminiscent of the "stab-in-the-back" theory by which right-wing Germans of the 1920s explained the loss of World War I). But, in the second phase, the finger is pointed in a different direction: "The problem with U.S. competitiveness," writes Tribus, "is the quality of U.S. management."

The reform proposals start from the assertion that the issue is not Japanese management versus American management, but good management versus bad management. "There is no black magic or economic judo involved," Tsurumi points out. Tribus concurs, pointing to some evidence that the problems are managerial, not cultural. The more complicated the process gets, the more decisively the Japanese win. And, as both authors point out, Japanese-owned plants in the United States operated by American workers show the same excellence as those in Japan.

Tribus and Tsurumi have built cases for reform that differ a bit in tone and details, but are alike in their essentials. First, become market driven. Concentrate on producing products that have what the customer perceives as quality, rather than what the engineer perceives as features that are fun to build. Second, concentrate technical effort on processes. Reduce their variability. This will set up a positive feedback process because the less the variability, the more easy it is to isolate, analyze, and learn from defects, eliminate them, and make the process yet more uniformly defect-free. Third, use the workers' heads. Replace management (which might be defined as "using the managers' heads to run the workers' hands") with leadership. Eliminate levels of hierarchy, and put responsibility for control over quality and process improvement at least partly in the workers' hands. They each tie it up in a slogan. Tribus: "management by quality"; Tsurumi: "adaptive corporate behavior."

Much of this sounds merely reasonable, not revolutionary. And indeed, most of America's giant corporations might accept two-thirds of it. But not point three, with its taint of workplace democracy. This cuts against virtually every "corporate culture" in the Fortune 500. Consider one example, a corporation considered not a reactionary dinosaur, but an example of enlightened management. It is fond of announcing that there are only three solutions to the competitive problem: automate, emigrate, or evaporate. When it entered the automation business, one of its workers remarked: "this company should not have any trouble making robots; it's been making them out of its workers for years." And its chief economist has stated: "one of the problems with replacing all our workers with robots is, if we have a depression, who will we lay off?"

Those remarks illustrate a tendency to view competitiveness as something to be achieved by management, engineers, and robots—not by workers, who are dismissed at best as tools, at worst encumbrances. It is a tendency, as Tsurumi puts it, to "treat human beings as disposable costs rather than renewable assets." Worker participation is approached by cosmetic measures, like issuing "the company is me" T-shirts, rather than by power sharing. Some of the authors recognize that even suggesting power sharing will not be well received. Tribus, for example, is a bit apologetic about intruding the idea of democracy in the workplace. "It is not a question of whether worker involvement is a good thing," he explains. "Worker involvement is a means to an end, not an end in itself." But he follows this with a more revolutionary thought: "This is the era of the knowledge worker; sharing power is the only way to get things done efficiently and smoothly in an ever-changing environment."

The word "only" is the key word. In the past, worker participation was regarded as a political demand detrimental to productivity. Its advocates could therefore be condemned as modern-day Luddites (19th–century Englishmen who broke up spinning frames to protest automation). The situation is reminiscent of the alleged "trade-off" between low cost and high quality. That too was accepted wisdom until Japanese manufacturers showed that the trade-off was a fiction, and that in a well organized factory, the high quality producer can also be the low cost producer. That was impressive. But a demonstration that the trade-off between democracy and productivity is a fiction, and that the democratic workplace can be the more productive workplace, would be revolutionary.

In promoting that revolution, the ideas presented by Tribus and Tsurumi are appealing. But to achieve revolution, appeal is not enough. There is an entrenched system in place to resist it. That entrenched system is not an anachronism defended by fools and knaves. It is indeed management driven rather than customer driven. It indeed emphasizes product rather than process. It indeed maintains rigid workplace hierarchies. But those characteristics are not arbitrary. They are evolutionary solutions to real problems, as the historian Alfred D. Chandler, of the Harvard Business School, has demonstrated in his books *Strategy and Structure* (1960) and *The Visible Hand* (1978). He shows that present methods of managing giant corporations evolved in response to practical needs to coordinate production and distribution, keep resources productively occupied, and manage growth.

That established system will not be easily overthrown. Perhaps it should not be overthrown at all. Its defenders can present some strong objections to the reformers' case. Let us consider just a few of them.

First, do the reform proposals indeed capture the Japanese secret (or, more accurately, the secret of good management, wherever practiced)? At first glance, that reform program does indeed embody much of the appeal of the Japanese innovation miracle. Evidence for this statement comes in the paper by Uenohara. He touches many of the same themes: flatter hierarchies; more responsibility and initiative left to the worker; emphasis on quality. He downplays cultural differences, offering few of them beyond appropriately enigmatic references to the sun and foothills of Mount Fuji.

He tells us that his company, NEC, simply applies to the problem of innovation more engineers more directly and spread more evenly than would a typical American company. NEC has more than two engineers for each million dollars of sales, substantially more than the figure for a comparable U.S. company, General Electric (about 1.2 engineers for each million dollars of sales). Unlike the typical practice in American companies, NEC's corporate R&D lab is attached to operations, rather than to corporate headquarters. American companies typically concentrate the best people at the corporate lab. NEC spreads them around. As Uenohara puts it: "Each link in the chain from basic research to development to production to marketing must be as strong as the next...Japanese industry makes sure that there are no weak links and no exaggeratedly strong ones—that would be inefficient."

If the Japanese advantage is simply the sum of a lot of commonsense policies like these, then the change can be incorporated in an unthreatening manner. This is indeed the case made by Robert H. Hayes of the Harvard Business School in his 1981 article in the *Harvard Business Review*, "How Japanese Factories Work." Hayes argues that "the Japanese have achieved their current level of manufacturing excellence mostly by doing simple things but doing them very well and slowly improving them all the time." He concedes the need for workers to work smarter, but tends to suggest this can be a result of indoctrination or managerial paternalism, not something to be achieved by power sharing.

A second objection can be made to the reform proposal: it is grounded on a historical myth. Tribus and Tsurumi present the myth this way: in the beginning, American companies were run by sturdy

Yankee technologists, not bottom-line squeezing financiers. But then the financiers took over. And they turned to a renegade engineer (another stab-in-the-back theory), Frederick Winslow Taylor, who told them what they wanted to hear: that the essence of manufacturing is using the heads of the managers to operate the hands of the workers. Then, they failed to listen to the efficiency experts who succeeded Taylor, failing especially to listen to an expert in quality control named W. Edwards Deming. So he went to Japan, and taught the Japanese how to beat the United States.

Anyone who sees that last paragraph as a caricature is invited to reread the chapters by Tribus and Tsurumi. Note as you do that they present little historical evidence of the unqualified triumph of Taylorism. Nor do they show that financial leaders, or engineering leaders, always do what is expected of them when they reach the president's post. In fact, many of them do not. At least one large multinational company (General Electric) did its most impressive pioneering in research, engineering, and invention during the presidency of a former shoe salesman (Charles A. Coffin, circa 1900), and has been engaged in its greatest splurge of mergers, acquisitions, and divestitures during the reign of a Ph.D. engineer (John F. Welch, Jr., the current CEO). Another company (General Motors) had an engineer president (Alfred Sloan) whose most impressive actions were in the area of corporate organization, and a financier president (Roger Smith, the current CEO) who led the company into such high tech fields as electronics and information systems. A third, Du Pont, has had an engineer-executive (Crawford Greenewalt) who kept it out of nuclear power, and a lawyer-executive (Irving Shapiro) who brought it into biotechnology. A version of history based on a duel between white-hatted engineers and black-hatted financiers may appeal to an audience of engineers, but is a shaky basis for understanding innovation.

A third objection. Tribus states: "concentrate on the quality of the process...the quality of the product will then take care of itself." But is this self-evident? Much of the past success of America's giant companies has been achieved by product improvement: by winning at the game of technology leapfrog. That is, they found attractive inventions that have been made by independents or smaller companies—for example, the steam turbine, the refrigerator, cellophane, the jet engine, or the CAT scanner—then used them to win at leapfrog; that is, they put their strong engineering force to work to devise the product improvements that, coupled with a strong marketing and service effort, would bring victory.

That worked for U.S. corporate giants for a while. But then they started playing leapfrog in solid state electronics with a bunch of high tech kangaroos—Texas Instruments, Fairchild, and Intel, for example—and the results were embarrassing. Then they started playing leapfrog with Japanese porcupines, and the results were painful.

But is the answer to abandon the game, or to get better at leaping? Even those Japanese porcupines are taking hopping lessons. As Uenohara puts it: "Technological innovation will be the key, and we must continuously motivate our engineers to accept the challenge of innovation." Product innovation is even more likely to be the winning game for a high-labor-cost economy such as the U.S.

A final point is not so much an objection as a question: where do the engineers fit in? Here Beckurts' chapter is especially helpful in reminding us that the engineer's role is in part determined by a broader technical culture. He points out, for example, that "engineering in Europe has always maintained a scientific attitude" and that "the European engineer's quest is for technical elegance, cost-effectiveness, and reliability in a product—in some cases even for a little too much perfection perhaps." He also points to the increase in university training at the expense of practical experience. He does not go on to say so, but those characteristics suggest that such engineers will not have much sympathy for the proposals presented in the other chapters. Much the same may be true of American engineers. Separated from non-university trained workers by a growing cultural gulf, and focusing on technical perfection, they tend to accept their roles in hierarchical organizations in which the engineer creates the instructions that go into the manager's head to operate the worker's hands. But engineers also pride themselves on their objectivity, and have as one of their goals high productivity. Suppose it can be objectively shown that more democratic workplace organization promotes rather than retards productivity? Might that bring the engineers over to the reformers' side?

To sum up, these chapters provide valuable views into the engineering management methods of Japan and Germany. They present proposals that, if adopted, would radically change the way technology-based industries are led. The reform proposals emphasize process rather than product orientation; customer driven management; and, most controversially, increasing reliance on the judgment and intelligence of the worker.

If I have dealt at length with objections to the reform programs, it is not to belittle them but to strengthen them. Receptiveness to reform will be followed soon by a phase of reaction. The reform proposals will have to be put more convincingly than they have been put here to survive that reaction. The reaction will occur because of the element of power sharing that is central to the program. Its entrenched opponents will argue that it is based on a superficial reading of both history and the present situation. They may succeed in dismissing it out of hand as a sentimental advocacy of workplace democracy. But the program, and especially its central assertion that worker participation and higher productivity are mutually reinforcing, deserves a more careful look. The proposals in these papers are not presented well enough to convince the skeptics, but they are suggestive and stimulating enough to deserve a consideration on their merits, not the steamroller of reaction.

Response to Wise–Myron Tribus

In presenting a very complex issue, it is sometimes necessary to compress arguments so severely that they are easily misinterpreted. Wise's comments suggest that the compression was overdone. The phrase (not slogan) "management by quality" is shorthand for a definite procedure to replace management by objectives. Just as Odiorne defined a specific way to approach management, so this phrase may be used as a shorthand for an approach that is now being adopted in many organizations worldwide. It approaches any problem by asking first, "What are we trying to do?" and then "What would it mean to do it with the highest quality?" This approach (with well-defined follow-on procedures) has been applied to such diverse activities as running a chain of newspapers or the acquisition of a company, each time with the generation of insight and the development of a plan of action that is easily seen to be superior to more conventional approaches.

As Wise points out, the biggest barrier to the adoption of better management is the reluctance of so many managers to share power. Since the building of the Pyramids, management has been a privilege and labor has been a commodity. These habits will not be easily overthrown. Indeed, my studies of six companies in Japan that won the Deming Prize show that in each case the managements changed because they had to. In each case, when I asked the chief executive officer "Why did you change?", the answer was the same: They were looking into the abyss. Ford did not just get a better idea. Ford Motor Company was forced into the new way to manage because it was confronted with unmistakable evidence that the old ways did not work in the new era.

What has to be faced by the managers of the Fortune 500 now is that the old mindset cannot develop the competitive enterprise in the new era. Wise is correct; this thrust will be resisted. We may well see the destruction of our society in the same way that the nobility in France brought on the revolution and the guillotine. Everything depends upon the forces that impinge on our managers. Wise seems to think that a different way to put the matter will make it acceptable. History suggests otherwise; people in privileged positions will always find reasons to conclude that for the better good they should retain their powers and privileges.

Wise cites Chandler's books, which were written before the evidence was all in place, as a justification for the existing system of management. It is true that the current methods of management evolved in response to certain forces. What is not to be taken as true is the conclusion that these methods were the *only* possible responses.

As I have written elsewhere, Taylor's contributions provided an extraordinary surge to our economy, and I am fully appreciative of what was accomplished. But as Taylor himself said in his historic testimony before the U.S. Congress, what was done in the name of "scientific management" went much too far in depriving us of the thinking ability of the workers. I did not attack financial people or propose engineers to replace them; what I do wish to attack is the practice of trying to manage a company on financial figures. These are *lagging indicators*, not useful for decision making, but only for after-the-fact evaluation. In rapidly changing times a manager must learn to read the *leading indicators* that can be found only by examining the *processes.*

My personal view is that the people who come to the top should have very broad experiences and, if possible, broad educational backgrounds. (This is why I spent a career in fighting for more liberal education for engineers, and more technical education for liberal arts majors.) It is a misreading of the matter to say that I see history as a duel between white-hatted technologists and black-hatted financial people. The spectacular growth of Xerox was led by Joe Wilson, a truly educated man and my own model of the best of all possible executives for whom to work.

Finally, Wise disregards my admonition to distinguish between the two thrusts required for competitiveness: a) Invention and innovation and b) Quality and productivity. It takes both thrusts to win. Americans have put their faith on invention and innovation and most of my Japanese colleagues seem to agree that this has been our strength. My argument is that one of these two thrusts is not enough; it takes both. The Japanese are now scrambling to improve their ability in the former; we have to improve in the latter.

If you pay enough attention to the design process and the innovation process, the product will take care of itself. I stand by that statement.

Part III

Institutional and Organizational Factors

The Contributors

Frank L. Huband Director, Electrical, Communications, and Systems Engineering Division, National Science Foundation, Washington, D.C.; Member, IEEE

James D. Meindl Vice President of Academic Affairs and Provost, Rensselaer Polytechnic Institute, Troy, New York; former Director of Laboratories, Center for Integrated Systems, Stanford University, Stanford, California; Fellow, IEEE

Gene W. Dalton (discussant) Professor of Organizational Behavior, Graduate School of Management, Brigham Young University, Provo, Utah

Lee Tom Perry (discussant) Assistant Professor of Organizational Behavior, Graduate School of Management, Brigham Young University, Provo, Utah

11/FEDERAL SUPPORT OF RESEARCH AND DEVELOPMENT FOR INDUSTRIAL COMPETITIVENESS

Frank L. Huband

Since World War II, new technology has been responsible for close to half—44 percent—of productivity gain in the United States. This contribution is far more than those of increased capital, better education, greater economies of scale, and improved resource allocation.

The federal government pays for roughly half of the approximately $100 billion the United States spends each year on research and development. It is involved to some extent in R&D related to almost all sectors of industrial activity.

However, the division of federal R&D between the civilian and defense sectors has changed dramatically over the last 25 years. In the 1960s, civilian R&D grew rapidly, and from the mid-1960s to 1980 the civilian and military research levels were about equal. But in the last five years, the balance has again shifted strongly toward defense. It is projected that for fiscal year 1987 almost three-fourths of the federal R&D effort will go into defense-related research. Thus, although the United States per capita research and development expenditures are comparable to or greater than those of our major trading partners, U.S. per capita civilian R&D expenditures have for many years been substantially less than those of both Germany and Japan.

Another factor is that for the first 10 or 20 years following World War II, military research was strongly relevant to the technologies of the civilian world, such as computers, semiconductors, and nuclear power, and this military research was at the leading edge of such research in the world. But today, most military aircraft research and development, for example, has no direct applicability to the civilian sector, which requires aircraft of quite a different character. And, in computers and semiconductors, it is the civilian sector research that is often at the leading edge.

A possible cause of this shift in leadership is the marked decline in the proportion of military funding devoted to basic research since the enactment of the Mansfield Amendment in 1971. The Mansfield Amendment required that research funded by the Department of Defense be directly related to the DOD mission. The current administration has substantially increased military spending in basic research as a percentage of civilian R&D, but even so, the military R&D in the basic research areas has declined over the last few years.

Deteriorating trade balance

For whatever reason, over the last decade our competitive record in international trade has not been very good. During the last several years our trade balance has deteriorated at an accelerating rate. The situation is worst in the older industries; there, we are in substantial deficit. But even for high technology products, we are slipping toward a deficit.

Another important indicator is that, over the last 15 years, productivity increased substantially faster in those countries that expend smaller proportions of their gross national product for military purposes. As the leader of the western world, the United States has carried a defense burden not approached by its allies, and it appears to have hurt our industrial productivity.

Japan, with a one percent defense burden, had a 2.8 percent improvement in productivity. The United States is at the bottom of the list, with almost seven percent of its gross domestic product devoted to the military and only a 0.3 percent improvement in productivity. I am referring to national economic productivity, which is the gross domestic product divided by the number of workers.

Even the mechanism by which products of federally funded university research are transmitted to industry is deteriorating. Historically, industry has obtained the results of this research largely through hiring Ph.D.s who had assisted in the research in their student days. Over the past several years, however, a larger and larger percentage of our Ph.D. students are not from the United States. For engineering, the proportion of all doctoral degrees awarded to nonresidents or noncitizens has increased to over 50 percent in the last several years.

Although many of these foreign students will remain in the United States, an increasing number are being paid by companies in their homeland to come to the United States to earn their doctorates and return home with their newly acquired skills and knowledge. For example, the chairman of the board of Daewoo Corp., a Korean conglomerate, has decided to send 100 of his employees to the United States to get Ph.D.s in high technology areas at a cost of over $4 million a year. One major innovation coming from this program, he said, could pay for the whole operation.

R&D initiatives

In the last few years there have been several initiatives taken to help solve these problems. The National Science Foundation has established a series of engineering research centers at academic institutions. The purpose of the centers is to offer research opportunities in areas critical to U.S. economic competitiveness, to enhance cross-disciplinary research, to upgrade engineering education, to develop mechanisms for expanded interactions between industry and university researchers, and to experiment with new ways of disseminating research results from universities to industry.

In 1985, NSF received over 140 proposals for centers requesting a total of $2.2 billion over a five-year period. Forty-six states were represented, plus the District of Columbia and Puerto Rico, and over 3,000 faculty members were listed as prospective participants.

Dozens of the proposals were clearly outstanding and worthy of funding. But available NSF funds allowed only six of them to be given an award, with a nine month total expenditure of less than $10 million.

In 1986, NSF received over 100 proposals requesting about $1.5 billion over five years. The NSF staff has completed visits to 15 sites, and a decision is expected in the next few months as to which will be selected for engineering research centers this year. We have about $22 million available for the first- and second-year awardees.

The engineering research centers have stimulated a favorable response from industry. Already, the first-year awardees have received commitments of almost $15 million from industry, more than the amount that NSF has put into the program to date.

A related initiative provided the Department of Defense this year with about $100 million for university research. Part of that money is intended for a similar class of research centers. Unfortunately, the tight budget that restricted the available funds for NSF has also hit the DOD program, and it appears that it will exist only for 1986.

A third way in which basic research in universities is being better tapped is by greater participation of the private sector. The Center for Integrated Systems, which James Meindl describes in Chapter 12, gives participating industrial firms access to the talents of Stanford University faculty members and their students. Another example is the Semiconductor Research Corporation, which funds academic research, much of it complementary to work funded by DOD and the National Science Foundation.

Finally, NSF's Presidential Young Investigator Awards to new faculty members encourage recipients to seek funds from industry by providing matching funds up to a nominal limit.

Taking larger steps

In the last year or so, there have been proposals for larger steps in government sponsorship of R&D to improve industrial competitiveness. The President's Commission on Industrial Competitiveness identified the U.S. science and technology knowledge base and U.S. talent as the country's only international competitive advantage. The report recommended expanded federal support for manufacturing research in universities.

The White House Science Council will soon publish the report of its Panel on the Health of U.S. Colleges and Universities. This document, already named the Packard Report after its chairman, David Packard, co-founder of Hewlett Packard, states that strong university-government-industry partnerships are fundamental to meeting our goals in economic competitiveness, and recommends that the federal government increase its investment in our institutions of higher education to enable them to grow with the demands of our society.

The National Science Foundation Advisory Council issued a report last fall (1985) which recommended realigning the goal structure in universities to accommodate activity in goal-directed research.

And last April, the National Academy of Engineering issued a report on the National Science Foundation's engineering program, recommending focused fundamental engineering research in emerging areas of technology and a tripling of the engineering directorate's budget to $400 million.

George Keyworth, when he was science advisor in the Office of Science and Technology Policy, aggressively promoted the idea of adding $500 million a year to our basic research funding, devoted entirely to starting up 50 science and technology centers. James McTeague, who is acting in Dr. Keyworth's old job, has recently restated strong support for the initiative to establish university-based multidisciplinary, problem-focused science and technology centers, aimed at areas of broad national needs and relevant to industrial technology. The centers' work would be general in nature, and not focused too closely on a given area, but ready to respond to opportunity.

Erich Bloch, the director of the National Science Foundation, has recommended another possible way to shift the resources of the United States toward the goal of developing a science and technology base for U.S. industrial competitiveness. He proposes that the bulk of the transferred resources be directed into the nation's universities. He points out that our universities have always attracted our best minds, and that universities combine research and education in a synergistic way that is impossible in other settings.

Mr. Bloch proposes that the funding for increased federal support of academic research come from a minor reallocation of the federal applied research and development accounts without any overall increase in the federal budget. He reports that a two percent reduction in those accounts would make $1 billion available. If these funds were used effectively, the result should be an overwhelming improvement in our overall rate of technological progress.

Nam Suh, who is assistant director of NSF for engineering, has proposed an even more aggressive initiative, which he calls Strategic Innovations for American Prosperity. This initiative would fund clusters of universities with the support of industrial participants, as appropriate, for a series of perhaps ten major comprehensive research programs in specific problem areas. Each program would have a budget of perhaps $50 million. Dr. Suh is identifying candidate general problem areas which might be best suited to this high intensity approach.

In order for any of these proposals or panel recommendations for major initiatives to come about, there must be a groundswell of support from just the mix of people represented by the IEEE. The federal government is entering into a period of such austerity that only initiatives with strong and almost unanimous support have any chance of seeing the light of day. The support of each major interest in the engineering community will be necessary if increased federal funding for basic and general research to improve competitiveness is to have a chance of success.

What if?

What if increased federal funds don't become available? There are still some steps that can make a difference. First, industry can help by increasing its participation in joint ventures and in cooperative research with the universities. The latter step serves the dual purpose of transferring to industry the insights acquired in academia and sensitizing the academic community to the research needs of industry.

Second, in light of the growing DOD research budget, it's important to find ways to exploit military R&D for civilian purposes. One small activity that NSF has been involved in is developing access to certain highly instrumented DOD basic research laboratories for academic research carried out by NSF-funded academic investigators. We have had excellent cooperation with respect to facilities, having used microelectronic laboratory equipment in extremely short supply for NSF researchers. If these experiments are as successful as I believe they will be, we plan to expand the concept to a broader range of laboratories.

A more difficult, but potentially powerful scheme is to make available to selected nondefense companies certain research results that could contribute to the companies' competitiveness. One way of providing this information is to conduct sessions similar to the current semiannual VHSIC contractor meetings, with the subject matter scrutinized to eliminate information that affects national security. Developing the rules for such access and selecting the participants would be difficult, but I believe the benefits would be worth the difficulty and risk and would be worth pursuing.

Research and development funded by the federal government does have a role to play in increasing U.S. industrial competitiveness. Whether or not new federal funds are made available for the task, industry, the universities, and the federal government will have to work toward these goals with greater vigor. The task is too important to be ignored. It is up to the leaders in these sectors to work together to restore the nation's industrial competitiveness.

12/ JOINT UNIVERSITY-INDUSTRY-GOVERNMENT RESEARCH TO FULFILL NATIONAL NEEDS

James D. Meindl

The Stanford University Center for Integrated Systems is a paradigm of joint university-industry-government research to meet an urgent national need. The urgent need is for improving the international competitiveness of United States industry. The Center seeks to help by developing computer-based manufacturing methods for ultra-large-scale integrated circuits (ULSI).

The principal intellectual goal of the Center is to blend two technological cultures: those of integrated electronics and of computer/communication systems. Integrated electronics is chiefly concerned with the fabrication processes and the transistor physics of semiconductor chips. Computer/communication systems are concerned with the software, hardware, and architecture of systems ranging from microcomputers to the largest supercomputers, and including voice, video, and data communication systems.

The Center for Integrated Systems is a cross-disciplinary unit within the Stanford School of Engineering. Its sponsors are several major U.S. corporations (Table 12-1). The sponsors built the Center's $15 million headquarters building and contribute $4 million annually for a core research program. To this amount, contracts and grants from the federal government add another $20 million per year. More than 80 Stanford faculty members are affiliated with the Center, and they supervise about 350 Ph.D. candidates. These students, who are really the brainpower of the Center, come chiefly from the Departments of Electrical Engineering, Mechanical Engineering, Chemical Engineering, Computer Science, Industrial Engineering and Engineering Management, and Materials Science and Engineering, and from the Graduate School of Business. In addition, each sponsoring company is invited to assign a member of its technical staff to work full time at the Center.

TABLE 12-1
LIST OF SPONSORS OF THE CENTER FOR INTEGRATED SYSTEMS

General Electric Company
Hewlett-Packard Company
TRW Incorporated
Northrop Corporation
Xerox Corporation
Texas Instruments Incorporated
Fairchild Camera and Instrument Corporation
Honeywell Incorporated
International Business Machines Corporation
Tektronix Incorporated
Digital Equipment Corporation
Intel Corporation
International Telephone and Telegraph Corporation
General Telephone and Electronics Corporation
Motorola Incorporated
United Technologies Corporation
Monsanto Industrial Chemical Company
Gould Incorporated/American Microsystems Incorporated
Philips Research Laboratories/Signetics Corporation
Rockwell International

The Center's laboratories in the headquarters building include 10,000 square feet of clean rooms, all with state-of-the-art control of temperature, humidity, vibration, and dust.

The Center's focus is on manufacturing. President Reagan's Commission on Industrial Competitiveness highlighted the crucial need for this focus when it stated "Perhaps the most glaring deficiency in America's technological capabilities has been our failure to devote enough attention to manufacturing or process technology [1]."

The Center's choice of ULSI as its target technology is a natural result of the historic growth of integrated circuit complexity (Fig. 12-1). In 1959, a silicon chip contained only a single transistor. By 1972, the number of transistors per chip had increased to approximately 8,000—equivalent to a doubling of complexity every year during the intervening period.

Since 1972, the complexity has quadrupled every three years, and it is now about five million transistors per chip. How long will the current exponential rate of increase continue? I have studied this question carefully, and have suggested that by the year 2000, silicon

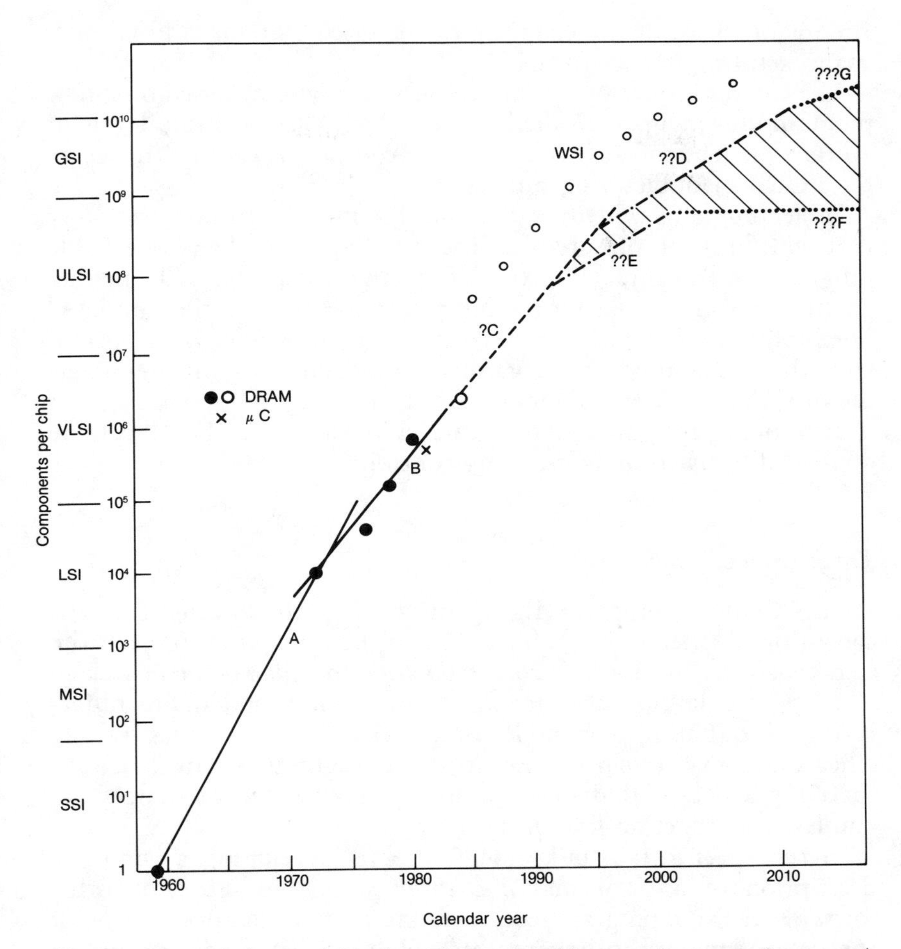

Fig. 12-1: *Growth in number of transistors per chip over the years. Segment A corresponds to an annual doubling, Segment B to a triannual quadrupling. Segments C through F are speculative projections. Wafer scale integration (WSI) may increase chip complexity by ten times in some applications such as memory.*

chips containing about one billion transistors each will be emerging from research and development laboratories [2].

Surely this ultra-large-scale integration in future computer, communication, control, and measurement systems will have a profound and pervasive effect all over the world. Just as surely, the United States cannot afford to have these systems, or their critical ULSI components, manufactured for it by foreign competitors. That is why the major thrust of the Center for Integrated Systems is to advance manufacturing science for ULSI.

The Center is developing a novel software system, based on a new programming language called Fable, which uses multiple levels of abstraction. The software is intended to completely capture the recipe for a ULSI manufacturing process.

Fable programs will directly control a manufacturing laboratory that will fabricate prototypical integrated systems. The same Fable programs will simulate a manufacturing line, and will predict parameter distributions for the integrated systems that the simulated line produces. Finally, Fable will compare the predicted distributions with those measured in laboratory production, and will iteratively modify the simulation model until it brings the two sets of distributions into close agreement. At that point, Fable can be applied directly to manufacturing control.

Three projects, nine tasks

The Center's program is organized into three projects, each consisting of three tasks (Table 12-2). In the integration project, the Center is carrying out a *factory modeling and management* task in which we are developing a statistical performance model of manufacturing throughput and cycle time. The model defines factory scheduling, maintenance, inspection, throughput, setup, and calibration practices so that an on-campus manufacturing laboratory can emulate a production line in industry.

In the *specification* task, the Center is developing a complete description of a ULSI manufacturing process in terms of Fable software. A Fable program receives as input an overall designation of the type of manufacturing process—NMOS, silicon-gate, 3-micron line width technology might be specified, for example (Fig. 12-2). In a first level of abstraction, with inverse device models, a Fable program and data base define the transistor and interconnection structures corresponding to the designated process. In a second level of abstraction, with inverse process models, the Fable software system defines the process conditions for creating the transistor and interconnection structures. In a third level of abstraction, this time with inverse equipment models, the Fable system defines the manufacturing-line control settings for the process conditions.

In the *simulation task*, we are developing a manufacturing-line simulator incorporating a concatenated set of circuit, device, process, and equipment models. The simulator is the quintessential requirement for establishing a science of ULSI manufacturing because it will

TABLE 12-2
MANUFACTURING SCIENCE PROGRAM ORGANIZATION AT THE CENTER FOR INTEGRATED SYSTEMS

Project	Task	Participants
Integration	Factory modeling and management	GSB, IEEM, I
	Specification	CS, EE, I
	Simulation	EE, I
Technology	Lithography	EE, ChE, I
	Etching	EE, ChE, ME, I
	Deposition	EE, ChE, ME, MS&E, I
Measurements and applications	Testing and diagnostics	EE, I
	Sensors and instrumentation	EE, ME, I
	System applications	EE, CS, I

Code: GBS = Graduate School of Business; IEEM = Department of Industrial Engineering and Engineering Management; I = industrial sponsor; CS = Department of Computer Science; EE = Department of Electrical Engineering; ChE = Department of Chemical Engineering; ME = Department of Mechanical Engineering; MS&E = Department of Materials and Science Engineering.

predict parameter distributions. Up until now, such simulators just haven't been available, primarily because suitable equipment models have been lacking, but also because software has been ineffective in combining available models for processes, devices, and circuits in a manufacturing environment. The importance of equipment models was not recognized until recently. An example of a useful equipment model is one that would predict the temperature, gas concentration, and gas flow at any point in space and time within the quartz tube of a resistance-heated oxidation furnace, given the control settings.

The goal of the technology project is to develop equipment models for the three general classes of equipment used in ULSI manufacturing. The models haven't been developed before because the U.S. companies that make semiconductor manufacturing equipment are predominantly small and can't afford the necessary research. Now, however, the Center proposes to join with these companies in creating the models. Thus, in the *lithography* task, we will create models for forming and transferring the patterns that define the lateral features of chips with beams of electrons, light, or x-rays. In the *etching* task, we will create models for removing material from a

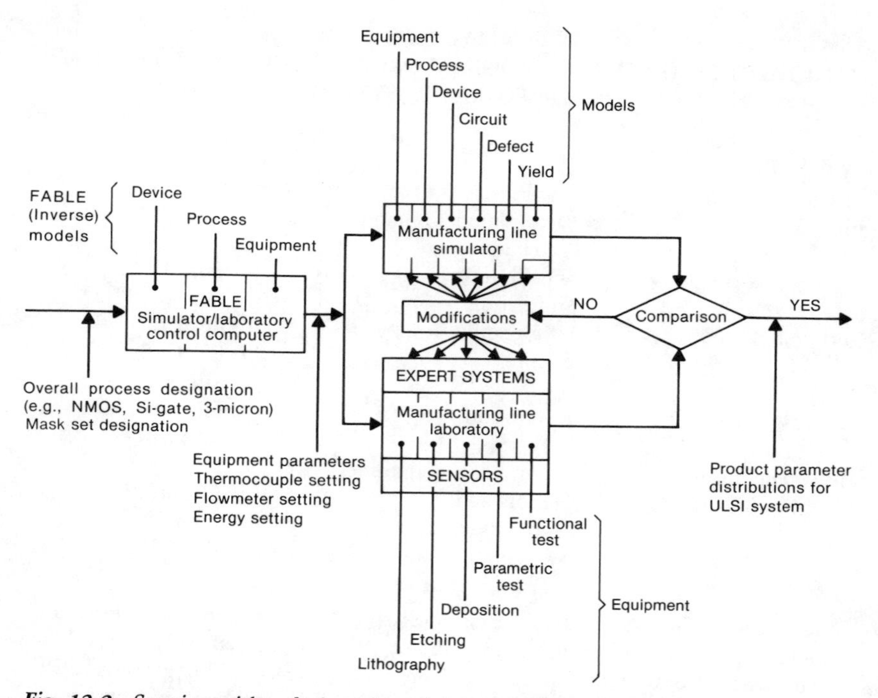

Fig. 12-2: Starting with a designation of the process technology, Fable software will generate product parameter distributions for a ULSI system.

semiconductor wafer with liquid or gaseous agents. And in the *deposition* task, we will create models for the addition and redistribution of materials in a wafer by such techniques as ion implantation, chemical vapor deposition, sputtering, and rapid thermal annealing.

The measurements and applications project is aimed at improving sensors and controls and demonstrating the performance of the manufacturing laboratory. The *testing and diagnostics* task will develop two classes of test chips to monitor the behavior of a manufacturing line. In-process test chips will provide fast measurements of the behavior of individual pieces of manufacturing equipment. End-of-process chips will be used to analyze the behavior of the complete manufacturing line. The end-of-process chips will include arrays of parametric test structures to pinpoint the densities of all types of defects introduced in manufacturing. They will provide a basis for defect and yield models in the manufacturing-line simulator.

The primary goal of the *sensors and instrumentation* task is to develop custom sensors for manufacturing equipment for purposes of modeling and controlling processes more accurately with micro expert systems. For example, a silicon wafer can be covered with an array of thin-film thermocouples and hard-wired to a laboratory computer. The wafer can then be used to map the temperature distribution in typical etching and deposition environments.

The *system applications* task will make it possible to demonstrate and verify the capabilities of the manufacturing laboratory through actual fabrication and functional testing of prototype integrated systems. In some cases, these systems will be designed with custom signal preprocessing circuitry to enhance the effectiveness of the manufacturing laboratory itself.

Potential payoffs

Stanford has done research on models for semiconductor devices and products for more than a decade. An important product of this work is the Suprem family of process simulators now in use at approximately 400 sites in U.S. industry. But the manufacturing modeling and control work now underway at the Center for Integrated Systems is at least 10 times more difficult than previous problems. Only radically new joint university-industry-government research efforts like that embodied in the Center can provide the necessary resources.

The potential rewards make the effort worthwhile:

- Manufacture of ULSI systems will be more repeatable because their recipes will be completely captured in Fable software.
- Manufacture of ULSI will be more flexible; it will provide application-specific fabrication capability to complement the design capability that exists now.
- Products can be designed for ease of manufacture in a systematic way with a manufacturing-line simulator.
- Turnaround in both design and manufacture will be faster.
- Manufacturing costs will be lower because of repeatable, flexible, high-yield, fast-turnaround fabrication.
- Manufacturing science for ULSI will become defined and established as a new intellectual discipline.
- The best and brightest engineering graduate students will be attracted to the field of manufacturing ULSI systems at the outset of their careers.

Beyond these benefits, the many small U.S. producers of semiconductor manufacturing equipment will have a national resource at their disposal. There are about 800 such companies, but less than 20 percent of them have sales greater than $20 million, and none has sales greater than $300 million. Nevertheless, it is a crucial industry, and joint research with universities and government will give it the leverage it needs to develop further.

Meanwhile, what can government do? A 1986 report by the White House Science Panel on the Health of U.S. Universities and Colleges made some cogent recommendations [3]. Industry should receive a 25 percent tax credit for support of university research, the panel said. Industry should receive a 100 percent tax deduction for donation of equipment to universities. Federal support of university research should be increased significantly. Implementation of these and other panel recommendations would strongly benefit research on national needs. Indeed, an even larger tax credit than 25 percent for support of university research should be given serious consideration.

References

[1] J. A. Young, "Global Competition—The New Reality," Report of President Reagan's Commission on Industrial Competitiveness, U.S. Govt. Printing Office, 0-481-213, Jan. 1985.

[2] J. D. Meindl, "Ultra-Large Scale Integration," *IEEE Trans. Electron Devices*, vol. ED-31, no. 11, pp. 1555–1561, Nov. 1984.

[3] K. McDonald, "Strength of U.S. Said to Depend on Universities," *The Chronicle of Higher Education*, vol. XXXI, no. 20, p. 1, Jan. 29, 1986.

13/ORGANIZING FOR INNOVATION

Gene W. Dalton and Lee Tom Perry

Mr. Huband, of the National Science Foundation, has written persuasively that increased financial support for university-based research and development is vital for the United States if we are to be industrially competitive. Mr. Meindl, of Stanford's Center for Integrated Systems, has described a paradigm for a joint university/industry/government program that has succeeded in attracting funding. Meindl also urges that if U.S. industry is to improve its international competitiveness, the government must provide significantly more funding for university research and tax incentives for industry to do likewise.

We would like to support their recommendations. Research with which we are familiar indicates that radical product innovations either do not originate in large industrial organizations, or, when they do originate within, are rejected by them [1]. Since it is these radical product innovations that initiate new product class life cycles, open up new relations with new vendors, cause extensive replacement of existing capital with new types of equipment, and establish new links with new scientific disciplines, this paucity of radical product innovation in large organizations is of major national concern. At the same time, it has been observed that radical product innovations occur in disproportionate numbers in companies and units located near strong, science-based universities [2]. We are not aware of comparable studies on process innovations, but our experience with radical product innovations and the potential link between product and process innovations, which we will discuss later, argue in favor of increasing the amount of research on products and processes in university settings.

Nevertheless, it is not enough to take note of the above-mentioned correlations, transfer significant amounts of funding to universities, and hope that past trends will continue. Innovation is as much an organizational issue as it is a technological or funding issue. Recently, we have been trying to understand the organizational processes that lead to radical product innovation in small and medium-sized firms that have their roots in university research. The papers by Huband and Meindl raise some critical issues that have surfaced as we have conducted our research about organizing for radical product innovation. Our intent is to add our perspective to those already presented by addressing the issues in terms of their organizational implications.

I. Universities must insist on maintaining their values and conventions. Organizations that wish to foster radical product innovation should become more like universities, not vice versa.

We have heard many arguments for increasing university funding for program research. Huband and Meindl present some of the more compelling of these arguments. We have also heard concerns, especially when private industry becomes the major source of such funding, about the values and the traditional organizational processes of the university being compromised.

Our concern about these relationships is more practical than philosophical. We are concerned that, in order to expand or extend the stream of funding from private industry, university research centers might be forced to adopt some of its practices. We fear that by so doing, university research centers may sacrifice their capacity for radical innovation.

We noted Abernathy's and Utterback's observation that radical product innovations occur in disproportionate numbers in companies and units located near strong science-based universities [3]. But it has been shown that it is not only the exchange of technology and people that is important to the symbiotic relationship between major research universities and companies dedicated to radical product innovation, it is also the exchange of "academic" values and conventions [4]. New start-ups that have been radically innovative have created organizational forms reflecting their university roots.

Let us illustrate this by describing the organizational form of one high tech firm we have been studying for more than five years. We shall call it Star Electronics. Star's fifteen year history is unusual in that gross sales have doubled roughly every two years surpassing $100 million annually, yet part of Star has retained its highly innovative form.

Initially, we were surprised by the organizational form that existed at Star. It appeared to violate numerous traditional management principles. The management philosophy and style of Star's founder and CEO particularly puzzled us. Then we realized that the organizational form the founder had created and nurtured at Star reflected the "academic" values of the university electrical engineering department he had left soon after he launched the company. Three characteristics of Star's organizational form seemed especially important to the continued fostering of radical product innovation, at least in this firm.

A. First assignments. Like many directors of research organizations, Star's CEO believes in recruiting the "best and the brightest"—in Star's case, the top five percent from the nation's premier electrical engineering schools. But unlike other research directors, he often gives these high-powered new hires job assignments that border on the banal. Why? Often, he claims, it is because he doesn't know where they will best fit in. Rather than predefine an area of contribution, he will assign a mundane task, then wait for the engineer to weary of it.

If the engineer is any good, the founder maintains, he will soon return with his own proposal for a more challenging assignment. If the proposal makes sense and relates (however tangentially) to Star's overall mission, the engineer is allowed to pursue it. If, on the other hand, the engineer continues to perform the mundane work, waiting for someone else to give him another assignment, then he is the "wrong type" for Star. He will not be terminated; instead, he will continue to receive low-level assignments until, frustrated, he leaves the company.

B. Articulation of organizational goals. While Star's founder intuitively knows where he wants the company to go in developing new technologies, he does not clearly communicate it. Part of this is due to the limitations of his intuitive vision—he only knows what he wants when he sees it—but it is more because he considers any sort of "imposed vision" inappropriate for his organization.

Some managers are frustrated by a perceived "lack of direction" at Star. They assume that the founder does not have a clear vision of where the company should go; that he is simply "muddling through." The founder's second in command explains, "[Star's founder] is a brilliant strategic thinker, but he never puts it down on paper. It's all in his head."

Fragments of Star's strategic focus are articulated to keep the company from veering too far off course. Often this occurs when a promising project that has been nurtured by a group of engineers is suddenly rejected by the founder because it does not conform to his vision. On these occasions it is obvious that Star's founder has a clear idea of the markets in which he is interested in competing and the kinds of technologies he wants to develop. But unlike many CEOs, he refuses to articulate his goals unless he has to. And even when he does, it is always in small pieces and to select groups.

C. Management style. The previous two characteristics of the founder's style suggest a nontraditional approach to management. Under this "academic" pattern of management, Star's engineers tend to band together in small, highly autonomous work groups. Beyond granting or withholding resources, and sponsoring stringent technical design reviews, upper management makes few attempts to control the groups. The reigning assumption is that the engineers should design and run their own projects. As a result, thick subcultures develop around these work groups. The groups become the primary socializing units, facilitating learning by apprenticeship. Group members put in long hours and harbor deep feelings about the importance of their work. The environment within these work groups has proved an ideal seedbed for the development of new technology.

We are not suggesting that all these characteristics must be present in technologically innovative organizations. What our research at Star Electronics does suggest is that radical product innovation requires very different organizational forms than exist in industry. It is difficult to imagine the CEO of a major corporation being tolerant of the conditions that exist at Star Electronics. He would perceive inefficiency, a lack of direction, and chaos. If such conditions existed at a university research center that had entered into a joint arrangement with his company, and he knew about them, he might work to change them. The representatives of government funding agencies, with their bias toward goal-directed, programmatic research, might take a similarly dim view. But from the perspective of Star's founder, such conditions are an inevitable by-product of the "academic" values and conventions that ensure radically innovative work. The dysfunctions could perhaps be minimized, but only at the cost of disrupting or stifling what had proven to be a remarkably liberating organizational form. More importantly, if the values and the organizational controls at the university itself were changed, the seedbed from which so much of our radical technical innovation springs could be spoiled.

II. The kind of funding for radical product innovation is as important as the amount of funding. Meindl and Huband, in their respective chapters, both argue for increased funding to university research centers. Our own research tells us that the *kind* of funding given in support of radical product innovation is equally important. Vast amounts of money are spent supporting large, integrated corporate R&D departments, but these units develop mostly "technologically-active middle range" innovations, not radical innovations [5]. On the other hand, developers of radical product innovations often have started working out of their homes to keep overhead costs to a minimum. They have relied more upon "sweat capital" than dollars for materials and equipment [6].

One high tech entrepreneur we interviewed suggested that the best kind of funding for radically innovative work comes in the form traditionally given by foundations. He explained that this was because the funding was tied to a general concept, not a specific research proposal. The funding was also secure and long-term in nature. Funding from foundations allowed him to retain enough control to pursue his long-term vision of an ultimate product technology.

In many of the high tech firms we have studied, an early indicator that the firm was moving away from radical product innovation occurred in the project review process. The emphasis shifted from reviewing the technical merits of a project to review of a business plan. Once this shift in emphasis occurred, the number of radical product proposals began to drop sharply. Radical product proposals could not be formatted as business plans without appearing both speculative and of unacceptably high risk.

The implication here is that the kinds of research proposals demanded by many funding agencies resemble business plans, especially in terms of the detail required. Radical product innovation often involves the exploration of "virgin" territory. It cannot be planned, except in a very general sense. If funding agencies wish to encourage radical innovation, they should, like foundations, fund *general concepts* and *people* more than detailed proposals.

III. The transfer of radical product innovations is as important as research and development. In many of the elite Japanese companies, engineers follow new technologies through the stages of research, design, development, and manufacturing. This is one of the reasons why Japanese firms, although not radically innovative, are highly successful at transferring technology. Technology transfer would be an easy matter if we could simply imitate the Japanese system. Unfortunately, and especially in the case of radical product innovation, engineers do not want to follow their new technologies very much beyond the design stage. Most of them want to return to basic research.

When both Meindl and Huband propose that we direct renewed emphasis and funding to university research centers, we must consider how we accomplish the transfer of technology out of the university environment. A few university researchers, of course, will take their new technologies out of the university themselves by starting new companies. Corporations can buy understanding of new technologies by hiring new Ph.D.s, as Huband observes. But these are both informal and somewhat arbitrary mechanisms for accomplishing the transfer of technology. And they may not work as well as they have in the past. This is clearly an area that requires more serious study and deliberate, imaginative experimentation.

IV. New advancements in process innovation create a need to rethink how we do radical product innovation. James Meindl's description of the process innovations coming out of Stanford's Center for Integrated Systems suggests that the relationship between product and process innovation is changing. Until very recently, an increased emphasis on process innovation meant that product innovation had become incremental. This was because a significant investment in process innovation would only occur once products were standardized [7]. But with the development of new process technologies with a previously unheard of degree of flexibility, past constraints on product innovation may no longer apply.

Several organizational scientists have conceptualized innovation in terms of a product life cycle. At one point in a product's life cycle a "dominant design" emerges that is an optimal product configuration. There is remarkable consensus among organizational scientists that the emergence of a "dominant design" is a positive, if not inevitable development in a firm's history. As Moore and Tushman note, the "dominant design is an extremely important event in the product life cycle...in that it fundamentally changes the nature of innovations, the bases of competition, manufacturing processes, and therefore marketing strategy and organizational design [8]." Our own research, however, clearly suggests that firms that want to continue to foster radical product innovation must retain an "open design." This means they can cater only to early adopters and must leave the market niches they have pioneered just as they become profitable in order to avoid head-to-head competition with more efficient producers.

If the promises of flexible process technologies, such as those being developed by the Center for Integrated Systems, are realized, it may soon be possible in some industries to retain an "open design," and also be an efficient producer. What this means is that institutions that foster radical product innovation will have the potential for receiving greater benefit from the technologies they create and develop. This should encourage more widespread radical product innovation. Moreover, it is also possible that flexible process technologies will become a direct source of radical product innovation because they make possible lower-cost experimentation.

Summary

It is clear to us that technological advancements must be coupled with serious investigation into their organizational implications. We have looked at issues such as the creation of an "academic" organizational form to foster radical innovation; kinds of funding; transfer of radical technologies; and the implications of new developments in process innovation to demonstrate the importance of organizational issues to the innovation process. Although we have avoided proposing specific strategies, our hope is that this discussion of issues will fuel further serious thinking about organizing for innovation.

References

[1] J. E. Goldman, "Innovation in Large Firms," in *Research on Technological Innovation, Management and Policy* Vol. 2, R. S. Rosenbloom, Ed. London: JAI Press, 1985, p. 1; W. L. Moore and M. L. Tushman, "Managing Innovation Over the Product Life Cycle," in *Readings in the Management of Innovation,* Moore and Tushman, Eds. Boston, MA: Pitman, 1982, pp. 133–34.

[2] W. J. Abernathy and J. M. Utterback, "Patterns of Industrial Innovation," *Technology Review,* vol. 80, p. 60, 1978.

[3] Ibid.

[4] S. Quince & Partners, *The Cambridge Phenomenon: The Growth of High Technology Industry in a University Town.* London: Brand Brothers and Company, 1985, p. 71.

[5] Abernathy and Utterback, p. 63.

[6] J. B. Quinn, "Technological Innovation, Entrepreneurship and Strategy," *Sloan Management Review,* pp. 21–22, Spring 1979.

[7] Abernathy and Utterback, pp. 59–61.

[8] Moore and Tushman, p. 134.

Part IV

Home and Abroad: Other Viewpoints

The Contributors

Robert H. Hayes Professor of Business Administration, Harvard Business School, Harvard University, Cambridge, Massachusetts
William J. Abernathy Professor of Business Administration, Harvard Business School, Harvard University, Cambridge, Massachusetts
Roland W. Schmitt Senior Vice President and Chief Scientist, General Electric Company, Schenectady, New York; Fellow, IEEE
Mark A. Fischetti Associate Managing Editor, *IEEE Spectrum*
Sir David Phillips Professor of Molecular Biophysics and Fellow of Corpus Christi College, University of Oxford, Oxford, England
Sir Hans Kornberg Professor of Biochemistry and Master of Christ's College, University of Cambridge, Cambridge, England
Derek H. Roberts Deputy Managing Director, General Electric Company, London; Member, IEEE
Daun Bhasavanich Senior Engineer, Plasma and Nuclear Science Department, Research and Development Center, Westinghouse Electric Corporation, Pittsburgh, Pennsylvania; Member, IEEE
Lawrence P. Grayson Advisor, Mathematics, Science, and Technology, Department of Education, Washington, D.C.; Fellow, IEEE

14/MANAGING OUR WAY TO ECONOMIC DECLINE

Robert H. Hayes and William J. Abernathy

During the past several years American business has experienced a marked deterioration of competitive vigor and a growing unease about its overall economic well-being. This decline in both health and confidence has been attributed by economists and business leaders to such factors as the rapacity of OPEC, deficiencies in government tax and monetary policies, and the proliferation of regulation. We find these explanations inadequate.

They do not explain, for example, why the rate of productivity growth in America has declined both absolutely and relative to that in Europe and Japan. Nor do they explain why in many high technology as well as mature industries America has lost its leadership position. Although a host of readily named forces—government regulation, inflation, monetary policy, tax laws, labor costs and constraints, fear of a capital shortage, the price of imported oil—have taken their toll on American business, pressures of this sort affect the economic climate abroad just as they do here.

A German executive, for example, will not be convinced by these explanations. Germany imports 95 percent of its oil (we import 50 percent), its government's share of gross domestic product is about 37 percent (ours is about 30 percent), and workers must be consulted on most major decisions. Yet Germany's rate of productivity growth has actually increased since 1970 and recently rose to more than four times ours. In France the situation is similar, yet today that country's productivity growth in manufacturing (despite current crises in steel and textiles) more than triples ours. No modern industrial nation is immune to the problems and pressures besetting U.S. business. Why then do we find a disproportionate loss of competitive vigor by U.S. companies?

Our experience suggests that, to an unprecedented degree, success in most industries today requires an organizational commitment to compete in the marketplace on technological grounds—that is, to compete over the long run by offering superior products. Yet, guided by what they took to be the newest and best principles of management, American managers have increasingly directed their attention elsewhere. These new principles, despite their sophistication and widespread usefulness, encourage a preference for (1) analytic detachment rather than the insight that comes from "hands on" experience and (2) short-term cost reduction rather than long-term development of technological competitiveness. It is this new managerial gospel, we feel, that has played a major role in undermining the vigor of American industry.

American management, especially in the two decades after World War II, was universally admired for its strikingly effective performance. But times change. An approach shaped and refined during stable decades may be ill suited to a world characterized by rapid and unpredictable change, scarce energy, global competition for markets, and a constant need for innovation. This is the world of the 1980s and, probably, the rest of this century.

The time is long overdue for earnest, objective self-analysis. What exactly have American managers been doing wrong? What are the critical weaknesses in the ways that they have managed the technological performance of their companies? What is the matter with the long-unquestioned assumptions on which they have based their managerial policies and practices?

A failure of management

In the past, American managers earned worldwide respect for their carefully planned yet highly aggressive action across three different time frames:

- *Short term*—using existing assets as efficiently as possible
- *Medium term*—replacing labor and other scarce resources with capital equipment
- *Long term*—developing new products and processes that open new markets or restructure old ones.

The first of these time frames demanded toughness, determination, and close attention to detail; the second, capital and the willingness to take sizable financial risks; the third, imagination and a certain amount of technological daring.

Our managers still earn generally high marks for their skill in improving short-term efficiency, but their counterparts in Europe and Japan have started to question America's entrepreneurial imagination and willingness to make risky long-term competitive investments. As one such observer remarked to us: "The U.S. companies in my industry act like banks. All they are interested in is return on investment and getting their money back. Sometimes they act as though they are more interested in buying other companies than they are in selling products to customers."

In fact, this curt diagnosis represents a growing body of opinion that openly charges American managers with competitive myopia: "Somehow or other, American business is losing confidence in itself and especially confidence in its future. Instead of meeting the challenge of the changing world, American business today is making small, short-term adjustments by cutting costs and by turning to the government for temporary relief... Success in trade is the result of patient and meticulous preparations, with a long period of market preparation before the rewards are available... To undertake such commitments is hardly in the interest of a manager who is concerned with his or her next quarterly earnings reports [1]."

More troubling still, American managers themselves often admit the charge with, at most, a rhetorical shrug of their shoulders. In established businesses, notes one senior vice president of research: "We understand how to market, we know the technology, and production problems are not extreme. Why risk money on new businesses when good, profitable low-risk opportunities are on every side?" Says another: "It's much more difficult to come up with a synthetic meat product than a lemon-lime cake mix. But you work on the lemon-lime cake mix because you know exactly what that return is going to be. A synthetic steak is going to take a lot longer, require a much bigger investment, and the risk of failure will be greater [2]."

These managers are not alone; they speak for many. Why, they ask, should they invest dollars that are hard to earn back when it is so easy—and so much less risky—to make money in other ways? Why ignore a ready-made situation in cake mixes for the deferred and far less certain prospects in synthetic steaks? Why shoulder the competitive risks of making better, more innovative products?

In our judgment, the assumptions underlying these questions are prime evidence of a broad managerial failure—a failure of both vision and leadership—that over time has eroded both the inclination and the capacity of U.S. companies to innovate.

Familiar excuses

About the facts themselves there can be little dispute. Tables 14-1–14-4 document our sorry decline. But the explanations and excuses commonly offered invite a good deal of comment.

TABLE 14-1
GROWTH IN LABOR PRODUCTIVITY SINCE 1960 (UNITED STATES AND ABROAD)

	Average annual percent change	
	Manufacturing 1960–1978	All industries 1960–1976
United States	2.8%	1.7%
United Kingdom	2.9	2.2
Canada	4.0	2.1
Germany	5.4	4.2
France	5.5	4.3
Italy	5.9	4.9
Belgium	6.9*	—
Netherlands	6.9*	—
Sweden	5.2	—
Japan	8.2	7.5

* 1960–1977.

Source: Council on Wage and Price Stability, *Report on Productivity.* Washington, D.C.: Executive Office of the President, July 1979.

It is important to recognize, first of all, that the problem is not new. It has been going on for at least 15 years. The rate of productivity growth in the private sector peaked in the mid-1960s. Nor is the problem confined to a few sectors of our economy; with a few exceptions, it permeates our entire economy. Expenditures on R&D by both business and government, as measured in constant (noninflated) dollars, also peaked in the mid-1960s—both in

TABLE 14-2
GROWTH OF LABOR PRODUCTIVITY BY SECTOR, 1948–1978

Time sector	Growth of labor productivity (annual average percent)		
	1948–65	1965–73	1973–78
Private business	3.2%	2.3%	1.1%
Agriculture, forestry, and fisheries	5.5	5.3	2.9
Mining	4.2	2.0	−4.0
Construction	2.9	−2.2	−1.8
Manufacturing	3.1	2.4	1.7
Durable goods	2.8	1.9	1.2
Nondurable goods	3.4	3.2	2.4
Transportation	3.3	2.9	0.9
Communication	5.5	4.8	7.1
Electric, gas, and sanitary services	6.2	4.0	0.1
Trade	2.7	3.0	0.4
Wholesale	3.1	3.9	0.2
Retail	2.4	2.3	0.8
Finance, insurance, and real estate	1.0	−0.3	1.4
Services	1.5	1.9	0.5
Government enterprises	−0.8	0.9	−0.7

Source: Bureau of Labor Statistics.

Note: Productivity data for services, construction, and finance, insurance, and real estate are unpublished.

TABLE 14-3
NATIONAL EXPENDITURES FOR PERFORMANCE OF R&D AS A PERCENT OF GNP BY COUNTRY, 1961–1978*

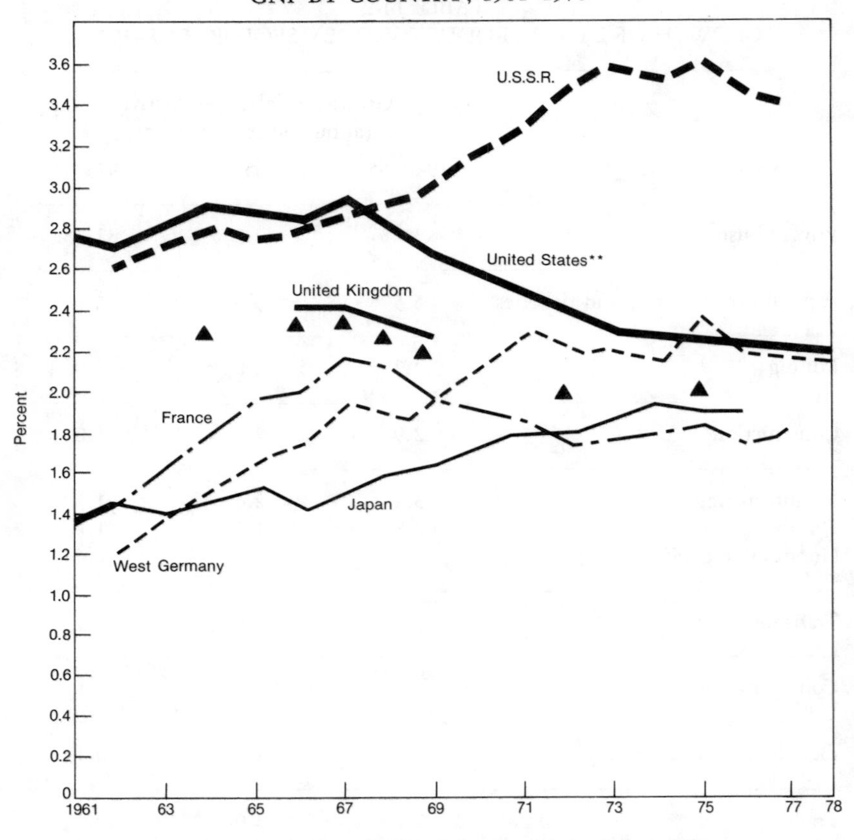

*Gross expenditures for performance of R&D including associated capital expenditures.
**Detailed information on capital expenditures for R&D is not available for the United States. Estimates for the period 1972–1977 show that their inclusion would have an impact of less than one-tenth of 1% for each year.
Note: The latest data may be preliminary or estimates.
Source: *Science Indicators-1978* Washington, D.C., National Science Foundation, 1979, p. 6

absolute terms and as a percentage of GNP. During the same period the expenditures on R&D by West Germany and Japan have been rising. More important, American spending on R&D as a percentage of sales in such critical research-intensive industries as machinery, professional and scientific instruments, chemicals, and aircraft had dropped by the mid-1970s to about half its level in the early 1960s. These are the very industries on which we now depend for the bulk of our manufactured exports.

Investment in plant and equipment in the United States displays

TABLE 14-4
U.S. INDUSTRIAL R&D EXPENDITURES

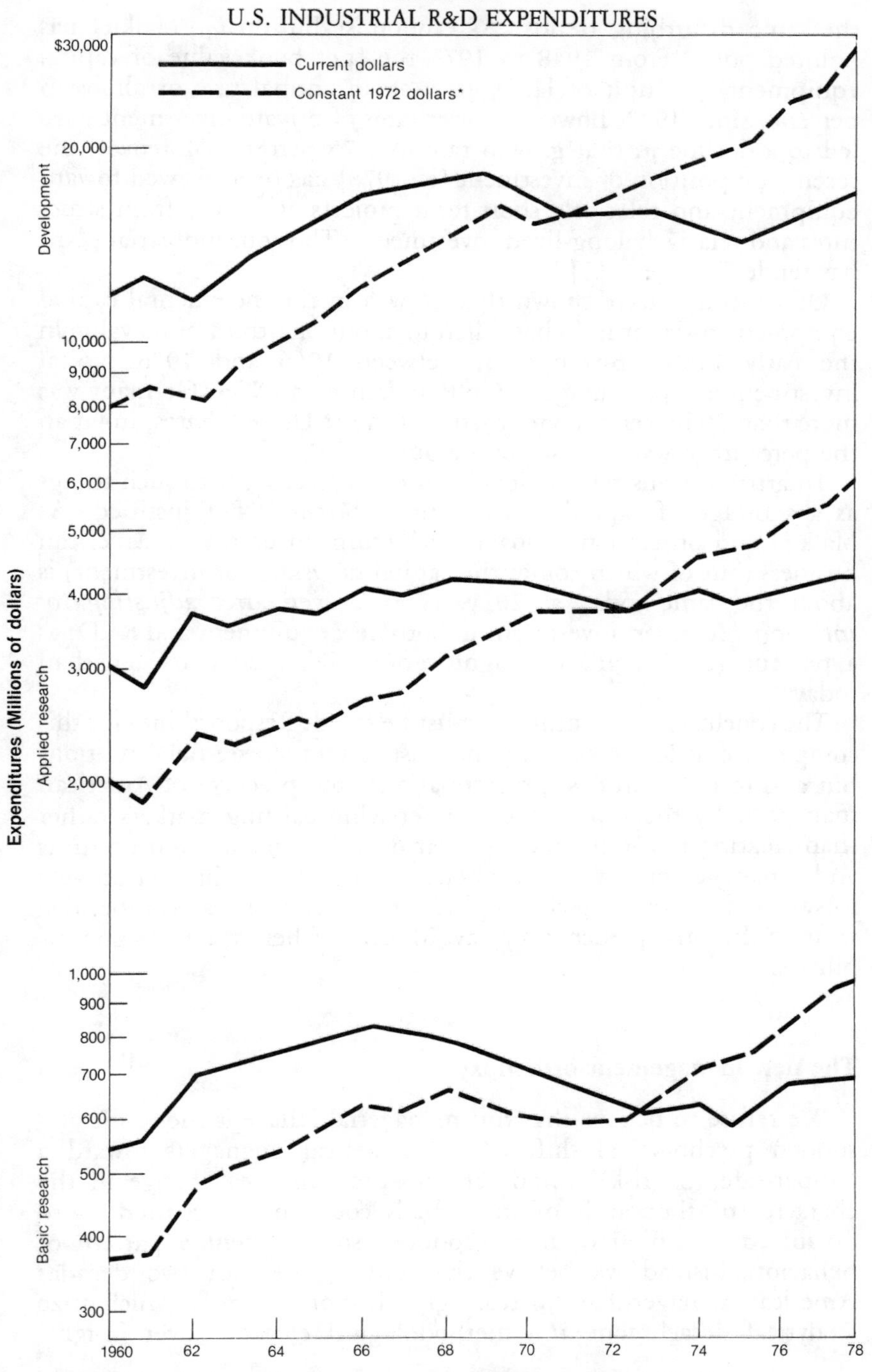

*GNP implicit price deflators used to convert current dollars to constant 1972 dollars.
Note: Preliminary data are shown for 1977 and estimates for 1978.
Source: *Science Indicators—1978*, p. 87

the same disturbing trends. As economist Burton G. Malkiel has pointed out: "From 1948 to 1973 the [net book value of capital equipment] per unit of labor grew at an annual rate of almost 3 percent. Since 1973, however, lower rates of private investment have led to a decline in that growth rate to 1.75 percent. Moreover, the recent composition of investment [in 1978] has been skewed toward equipment and relatively short-term projects and away from structures and relatively long-lived investments. Thus our industrial plant has tended to age... [3]."

Other studies have shown that growth in the incremental capital equipment-to-labor ratio has fallen to about one-third of its value in the early 1960s. By contrast, between 1966 and 1976 capital investment as a percentage of GNP in France and West Germany was more than 20 percent greater than that in the United States; in Japan the percentage was almost double ours.

To attribute this relative loss of technological vigor to such things as a shortage of capital in the United States is not justified. As Malkiel and others have shown, the return on equity of American business (out of which comes the capital necessary for investment) is about the same today as 20 years ago, *even after adjusting for inflation*. However, investment in both new equipment and R&D, as a percentage of GNP, was significantly higher 20 years ago than today.

The conclusion is painful but must be faced. Responsibility for this competitive listlessness belongs not just to a set of external conditions but also to the attitudes, preoccupations, and practices of American managers. By their preference for servicing existing markets rather than creating new ones and by their devotion to short-term returns and "management by the numbers," many of them have effectively forsworn long-term technological superiority as a competitive weapon. In consequence, they have abdicated their strategic responsibilities.

The new management orthodoxy

We refuse to believe that this managerial failure is the result of a sudden psychological shift among American managers toward a "super-safe, no risk" mind set. No profound sea change in the character of thousands of individuals could have occurred in so organized a fashion or have produced so consistent a pattern of behavior. Instead we believe that during the past two decades American managers have increasingly relied on principles which prize analytical detachment and methodological elegance over insight, based on experience, into the subtleties and complexities of strategic decisions. As a result, maximum short-term financial returns have become the overriding criteria for many companies.

For purposes of discussion, we may divide this *new* management orthodoxy into three general categories: financial control, corporate portfolio management, and market-driven behavior.

Financial control

As more companies decentralize their organizational structures, they tend to fix on profit centers as the primary unit of managerial responsibility. This development necessitates, in turn, greater dependence on short-term financial measurements like return on investment (ROI) for evaluating the performance of individual managers and management groups. Increasing the structural distance between those entrusted with exploiting actual competitive opportunities and those who must judge the quality of their work virtually guarantees reliance on objectively quantifiable short-term criteria.

Although innovation, the lifeblood of any vital enterprise, is best encouraged by an environment that does not unduly penalize failure, the predictable result of relying too heavily on short-term financial measures—a sort of managerial remote control—is an environment in which no one feels he or she can afford a failure or even a momentary dip in the bottom line.

Corporate portfolio management

This preoccupation with control draws support from modern theories of financial portfolio management. Originally developed to help balance the overall risk and return of stock and bond portfolios, these principles have been applied increasingly to the creation and management of corporate portfolios—that is, a cluster of companies and product lines assembled through various modes of diversification under a single corporate umbrella. When applied by a remote group of dispassionate experts primarily concerned with finance and control and lacking hands-on experience, the analytic formulas of portfolio theory push managers even further toward an extreme of caution in allocating resources.

"Especially in large organizations," reports one manager, "we are observing an increase in management behavior which I would regard as excessively cautious, even passive; certainly overanalytical; and, in general, characterized by a studied unwillingness to assume responsibility and even reasonable risk."

Market-driven behavior

In the past 20 years, American companies have perhaps learned too well a lesson they had long been inclined to ignore: businesses should be customer oriented rather than product oriented. Henry Ford's famous dictum that the public could have any color automobile it wished as long as the color was black has since given way to its philosophical opposite: "We have got to stop marketing makeable products and learn to make marketable products."

At last, however, the dangers of too much reliance on this philosophy are becoming apparent. As two Canadian researchers have put it: "Inventors, scientists, engineers, and academics, in the normal pursuit of scientific knowledge, gave the world in recent times the laser, xerography, instant photography, and the transistor. In contrast, worshipers of the marketing concept have bestowed upon mankind such products as new-fangled potato chips, feminine hygiene deodorant, and the pet rock... [4]."

The argument that no new product ought to be introduced without managers undertaking a market analysis is common sense. But the argument that consumer analysis and formal market surveys should dominate other considerations when allocating resources to product development is untenable. It may be useful to remember that the initial market estimate for computers in 1945 projected total worldwide sales of only ten units. Similarly, even the most carefully researched analysis of consumer preferences for gas-guzzling cars in an era of gasoline abundance offers little useful guidance to today's automobile manufacturers in making wise product investment decisions. Customers may know what their needs are, but they often define those needs in terms of existing products, processes, markets, and prices.

Deferring to a market-driven strategy without paying attention to its limitations is, quite possibly, opting for customer satisfaction and lower risk in the short run at the expense of superior products in the future. Satisfied customers are critically important, of course, but not if the strategy for creating them is responsible as well for unnecessary product proliferation, inflated costs, unfocused diversification, and a lagging commitment to new technology and new capital equipment.

Three managerial decisions

These are serious charges to make. But the unpleasant fact of the matter is that, however useful these new principles may have been initially, if carried too far they are bad for U.S. business. Consider, for example, their effect on three major kinds of choices regularly faced by corporate managers: the decision between imitative and innovative product design, the decision to integrate backward, and the decision to invest in process development.

Imitative vs. innovative product design

A market-driven strategy requires new product ideas to flow from detailed market analysis or, at least, to be extensively tested for consumer reaction before actual introduction. It is no secret that these requirements add significant delays and costs to the introduction of new products. It is less well known that they also predispose managers toward developing products for existing markets and toward product designs of an imitative rather than an innovative nature. There is increasing evidence that market-driven strategies tend, over time, to dampen the general level of innovation in new product decisions.

Confronted with the choice between innovation and imitation, managers typically ask whether the marketplace shows any consistent preference for innovative products. If so, the additional funding they require may be economically justified; if not, those funds can more properly go to advertising, promoting, or reducing the prices of less advanced products. Though the temptation to allocate resources so as to strengthen performance in existing products and markets is often irresistible, recent studies by J. Hugh Davidson and others confirm the strong market attractiveness of innovative products [5].

Nonetheless, managers having to decide between innovative and imitative product design face a difficult series of marketing-related trade-offs. Table 14-5 summarizes these trade-offs.

By its very nature, innovative design is, as Joseph Schumpeter observed a long time ago, initially destructive of capital—whether in the form of labor skills, management systems, technological processes, or capital equipment. It tends to make obsolete existing investments in both marketing and manufacturing organizations. For the managers concerned it represents the choice of uncertainty (about economic returns, timing, etc.) over relative predictability, exchanging the reasonable expectation of current income against the promise of high future value. It is the choice of the gambler, the person willing to risk much to gain even more.

Conditioned by a market-driven strategy and held closely to account by a "results now" ROI-oriented control system, American managers have increasingly refused to take the chance on innovative product/market development. As one of them confesses: "In the last

TABLE 14-5
TRADE-OFFS BETWEEN INITIATIVE AND INNOVATIVE DESIGN FOR AN ESTABLISHED PRODUCT LINE

Imitative design	Innovative design
Market demand is relatively well known and predictable.	Potentially large but unpredictable demand; the risk of a flop is also large.
Market recognition and acceptance are rapid.	Market acceptance may be slow initially, but but the imitative response of competitors may also be slowed.
Readily adaptable to existing market, sales, and distribution policies.	May require unique, tailored marketing distribution and sales policies to educate customers or because of special repair and warranty problems.
Fits with existing market segmentation and product policies.	Demand may cut across traditional marketing segments, disrupting divisional responsibilities and cannibalizing other products.

year, on the basis of high capital risk, I turned down new products at a rate at least twice what I did a year ago. But in every case I tell my people to go back and bring me some new product ideas [6].'' In truth, they have learned caution so well that many are in danger of forgetting that market-driven, follow-the-leader companies usually end up following the rest of the pack as well.

Backward integration

Sometimes the problem for managers is not their reluctance to take action and make investments but that, when they do so, their action has the unintended result of reinforcing the status quo. In deciding to integrate backward because of apparent short-term rewards, managers often restrict their ability to strike out in innovative directions in the future.

Consider, for example, the case of a manufacturer who purchases a major component from an outside company. Static analysis of production economies may very well show that backward integration offers rather substantial cost benefits. Eliminating certain purchasing and marketing functions, centralizing overhead, pooling R&D efforts and resources, coordinating design and production of both product and component, reducing uncertainty over design changes, allowing for the use of more specialized equipment and labor skills—in all these ways and more, backward integration holds out to management the promise of significant short-term increases in ROI.

These efficiencies may be achieved by companies with commodity-like products. In such industries as ferrous and nonferrous metals or petroleum, backward integration toward raw materials and supplies tends to have a strong, positive effect on profits. However, the situation is markedly different for companies in more technologically active industries. Where there is considerable exposure to rapid technological advances, the promised value of backward integration becomes problematic. It may provide a quick, short-term boost to ROI figures in the next annual report, but it may also paralyze the long-term ability of a company to keep on top of technological change.

The real competitive threats to technologically active companies arise less from changes in ultimate consumer preference than from abrupt shifts in component technologies, raw materials, or production processes. Hence those managers whose attention is too firmly directed toward the marketplace and near-term profits may suddenly discover that their decision to make rather than buy important parts has locked their companies into an outdated technology.

Further, as supply channels and manufacturing operations become more systematized, the benefits from attempts to "rationalize" production may well be accompanied by unanticipated side effects. For instance, a company may find itself shut off from the R&D efforts of various independent suppliers by becoming their competitor. Similarly, the commitment of time and resources needed to master technology back up the channel of supply may distract a company from doing its own job well. Such was the fate of Bowmar, the pocket calculator pioneer, whose attempt to integrate backward into semiconductor production so consumed management attention that final assembly of the calculators, its core business, did not get the required resources.

Long-term contracts and long-term relationships with suppliers can achieve many of the same cost benefits as backward integration without calling into question a company's ability to innovate or respond to innovation. European automobile manufacturers, for example, have typically chosen to rely on their suppliers in this way; American companies have followed the path of backward integration. The resulting trade-offs between production efficiencies and innovative flexibility should offer a stern warning to those American managers too easily beguiled by the lure of short-term ROI improvement. A case in point: the U.S. auto industry's huge investment in automating the manufacture of cast-iron brake drums probably delayed by more than five years its transition to disc brakes.

Process development

In an era of management by the numbers, many American managers—especially in mature industries—are reluctant to invest heavily in the development of new manufacturing processes. When asked to explain their reluctance, they tend to respond in fairly predictable ways. "We can't afford to design new capital equipment for just our own manufacturing needs" is one frequent answer. So is: "The capital equipment producers do a much better job, and they can amortize their development costs over sales to many companies." Perhaps most common is: "Let the others experiment in manufacturing; we can learn from their mistakes and do it better."

Each of these comments rests on the assumption that essential advances in process technology can be appropriated more easily through equipment purchase than through in-house equipment design and development. Our extensive conversations with the managers of European (primarily German) technology-based companies have convinced us that this assumption is not as widely shared abroad as in the United States. Virtually across the board, the European managers impressed us with their strong commitment to increasing market share through internal development of advanced process technology—even when their suppliers were highly responsive to technological advances.

By contrast, American managers tend to restrict investments in process development to only those items likely to reduce costs in the short run. Not all are happy with this. As one disgruntled executive told us: "For too long U.S. managers have been taught to set low priorities on mechanization projects, so that eventually divestment appears to be the best way out of manufacturing difficulties. Why?

"The drive for short-term success has prevented managers from looking thoroughly into the matter of special manufacturing equipment, which has to be invented, developed, tested, redesigned, reproduced, improved, and so on. That's a long process, which needs experienced, knowledgeable, and dedicated people who stick to their jobs over a considerable period of time. Merely buying new equipment (even if it is possible) does not often give the company any advantage over competitors."

We agree. Most American managers seem to forget that, even if they produce new products with their existing process technology (the same "cookie cutter" everyone else can buy), their competitors will face a relatively short lead time for introducing similar products. And as Eric von Hipple's studies of industrial innovation show, the innovations on which new industrial equipment is based usually originate with the user of the equipment and not with the equipment producer [7]. In other words, companies can make products more profitable by investing in the development of their own process technology. Proprietary processes are every bit as formidable competitive weapons as proprietary products.

The American managerial ideal

Two very important questions remain to be asked: (1) Why should so many American managers have shifted so strongly to this new managerial orthodoxy? and (2) Why are they not more deeply bothered by the ill effects of those principles on the long-term technological competitiveness of their companies? To answer the first question, we must take a look at the changing career patterns of American managers during the past quarter century; to answer the second, we must understand the way in which they have come to regard their professional roles and responsibilities as managers.

The road to the top

During the past 25 years the American manager's road to the top has changed significantly. No longer does the typical career, threading sinuously up and through a corporation with stops in several functional areas, provide future top executives with intimate hands-on knowledge of the company's technologies, customers, and suppliers.

Table 14-6 summarizes the currently available data on the shift in functional background of newly appointed presidents of the 100 largest U.S. corporations. The immediate significance of these figures is clear. Since the mid-1950s there has been a rather substantial increase in the percentage of new company presidents whose primary interests and expertise lie in the financial and legal areas and not in production. In the view of C. Jackson Grayson, president of the American Productivity Center, American management has for 20 years "coasted off the great R&D gains made during World War II, and constantly rewarded executives from the marketing, financial, and legal sides of the business while it ignored the production men.

TABLE 14-6

CHANGES IN THE PROFESSIONAL ORIGINS OF CORPORATE PRESIDENTS

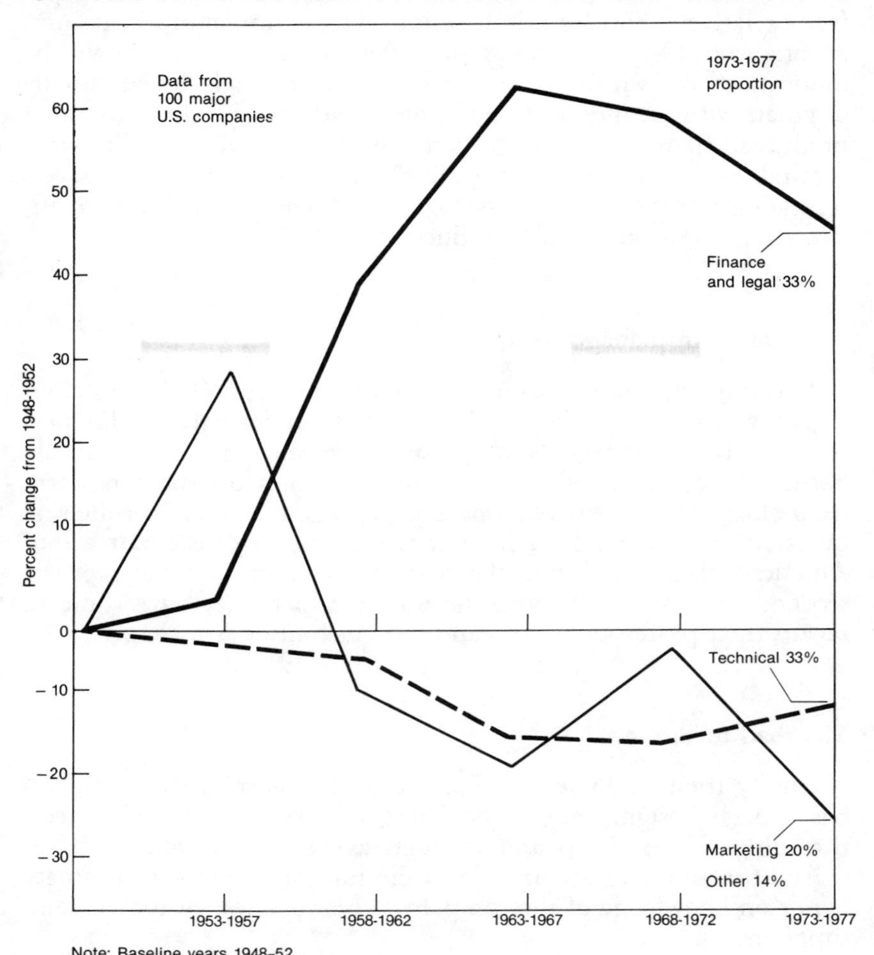

Note: Baseline years 1948–52
Source: Golightly & Co. International (1978)

Today [in business schools] courses in the production area are almost nonexistent [8].''

In addition, companies are increasingly choosing to fill new top management posts from outside their own ranks. In the opinion of foreign observers, who are still accustomed to long-term careers in the same company or division, ''High-level American executives...seem to come and go and switch around as if playing a game of musical chairs at an Alice in Wonderland tea party.''

Far more important, however, than any absolute change in numbers is the shift in the general sense of what an aspiring manager has to be "smart about" to make it to the top. More important still is the broad change in attitude such trends both encourage and express. What has developed, in the business community as in academia, is a preoccupation with a false and shallow concept of the professional manager, a "pseudo-professional" really—an individual having no special expertise in any particular industry or technology who nevertheless can step into an unfamiliar company and run it successfully through strict application of financial controls, portfolio concepts, and a market-driven strategy.

The gospel of pseudo-professionalism

In recent years, this idealization of pseudo-professionalism has taken on something of the quality of a corporate religion. Its first doctrine, appropriately enough, is that neither industry experience nor hands-on technological expertise counts for very much. At one level, of course, this doctrine helps to salve the conscience of those who lack them. At another, more disturbing level it encourages the faithful to make decisions about technological matters simply as if they were adjuncts to finance or marketing decisions. We do not believe that the technological issues facing managers today can be meaningfully addressed without taking into account marketing or financial considerations; on the other hand, neither can they be resolved with the same methodologies applied to these other fields.

Complex modern technology has its own inner logic and developmental imperatives. To treat it as if it were something else—no matter how comfortable one is with that other kind of data—is to base a competitive business on a two-legged stool, which must, no matter how excellent the balancing act, inevitably fall to the ground.

More disturbing still, true believers keep the faith on a day-to-day basis by insisting that as issues rise up the managerial hierarchy for decision they be progressively distilled into easily quantifiable terms. One European manager, in recounting to us his experiences in a joint venture with an American company, recalled with exasperation that "U.S. managers want everything to be simple. But sometimes business situations are not simple, and they cannot be divided up or looked at in such a way that they become simple. They are messy, and one must try to understand all the facets. This appears to be alien to the American mentality."

The purpose of good organizational design, of course, is to divide responsibilities in such a way that individuals have relatively easy tasks to perform. But then these differentiated responsibilities must be pulled together by sophisticated, broadly gauged integrators at the top of the managerial pyramid. If these individuals are interested in but one or two aspects of the total competitive picture, if their training includes a very narrow exposure to the range of functional specialties, if—worst of all—they are devoted simplifiers themselves, who will do the necessary integration? Who will attempt to resolve complicated issues rather than try to uncomplicate them artificially? At the strategic level there are no such things as pure production problems, pure financial problems, or pure marketing problems.

Merger mania

When executive suites are dominated by people with financial and legal skills, it is not surprising that top management should increasingly allocate time and energy to such concerns as cash management and the whole process of corporate acquisitions and mergers. This is indeed what has happened. In 1978 alone there were some 80 mergers involving companies with assets in excess of $100 million each; in 1979 there were almost 100. This represents roughly $20 billion in transfers of large companies from one owner to another—two thirds of the total amount spent on R&D by American industry.

In 1978 *Business Week* ran a cover story on cash management in which it stated that "the 400 largest U.S. companies together have more than $60 billion in cash—almost triple the amount they had at the beginning of the 1970s." The article also described the increasing attention devoted to—and the sophisticated and exotic techniques used for—managing this cash hoard.

There are perfectly good reasons for this flurry of activity. It is entirely natural for financially (or legally) trained managers to concentrate on essentially financial (or legal) activities. It is also natural for managers who subscribe to the portfolio "law of large numbers" to seek to reduce total corporate risk by parceling it out among a sufficiently large number of separate product lines, businesses, or technologies. Under certain conditions it may very well make good economic sense to buy rather than build new plants or modernize existing ones. Mergers are obviously an exciting game; they tend to produce fairly quick and decisive results, and they offer the kind of public recognition that helps careers along. Who can doubt the appeal of the titles awarded by the financial community; being called a "gunslinger," "white knight," or "raider" can quicken anyone's blood.

Unfortunately, the general American penchant for separating and simplifying has tended to encourage a diversification away from core technologies and markets to a much greater degree than is true in Europe or Japan. U.S. managers appear to have an inordinate faith in the portfolio law of large numbers—that is, by amassing enough product lines, technologies, and businesses, one will be cushioned against the random setbacks that occur in life. This might be true for portfolios of stocks and bonds, where there is considerable evidence that setbacks *are* random. Businesses, however, are subject not only to random setbacks such as strikes and shortages but also to carefully orchestrated attacks by competitors, who focus all their resources and energies on one set of activities.

Worse, the great bulk of this merger activity appears to have been absolutely wasted in terms of generating economic benefits for stockholders. Acquisition experts do not necessarily make good managers. Nor can they increase the value of their shares by merging two companies any better than their shareholders could do individually by buying shares of the acquired company on the open market (at a price usually below that required for a takeover attempt).

There appears to be a growing recognition of this fact. A number of U.S. companies are now divesting themselves of previously acquired companies; others are proposing to break themselves up into relatively independent entities. The establishment of a strong competitive position through in-house technological superiority is by nature a long, arduous, and often unglamorous task. But it is what keeps a business vigorous and competitive.

The European example

Gaining competitive success through technological superiority is a skill much valued by the seasoned European (and Japanese) managers with whom we talked. Although we were able to locate few hard statistics on their actual practice, our extensive investigations of more than 20 companies convinced us that European managers do indeed tend to differ significantly from their American counterparts. In fact, we found that many of them were able to articulate these differences quite clearly.

In the first place, European managers think themselves more pointedly concerned with how to survive over the long run under intensely competitive conditions. Few markets, of course, generate price competition as fierce as in the United States, but European companies face the remorseless necessity of exporting to other national markets or perishing.

The figures here are startling: manufactured product exports represent more than 35 percent of total manufacturing sales in France and Germany and nearly 60 percent in the Benelux countries, as against not quite 10 percent in the United States. In these export markets, moreover, European products must hold their own against "world class" competitors, lower-priced products from developing countries, and American products selling at attractive devalued dollar prices. To survive this competitive squeeze, European managers feel they must place central emphasis on producing technologically superior products.

Further, the kinds of pressures from European labor unions and national governments virtually force them to take a consistently long-term view in decision making. German managers, for example, must negotiate major decisions at the plant level with worker-dominated works councils; in turn, these decisions are subject to review by supervisory boards (roughly equivalent to American boards of directors), half of whose membership is worker elected. Together with strict national legislation, the pervasive influence of labor unions makes it extremely difficult to change employment levels or production locations. Not surprisingly, labor costs in Northern Europe have more than doubled in the past decade and are now the highest in the world.

To be successful in this environment of strictly constrained options, European managers feel they must employ a decision-making apparatus that grinds very fine—and very deliberately. They must simply outthink and outmanage their competitors. Now, American managers also have their strategic options hedged about by all kinds of restrictions. But those restrictions have not yet made them as conscious as their European counterparts of the long-term implications of their day-to-day decisions.

As a result, the Europeans see themselves as investing more heavily in cutting-edge technology than the Americans. More often than not, this investment is made to create new product opportunities in advance of consumer demand and not merely in response to market-driven strategy. In case after case, we found the Europeans striving to develop the products and process capabilities with which to lead markets and not simply responding to the current demands of the marketplace. Moreover, in doing this they seem less inclined to integrate backward and more likely to seek maximum leverage from stable, long-term relationships with suppliers.

Having never lost sight of the need to be technologically competitive over the long run, European and Japanese managers are extremely careful to make the necessary arrangements and investments today. And their daily concern with the rather basic issue of long-term survival adds perspective to such matters as short-term ROI or rate of growth. The time line by which they manage is long, and it has made them painstakingly attentive to the means for keeping their companies technologically competitive. Of course they pay attention to the numbers. Their profit margins are usually lower than ours, their debt ratios higher. Every tenth of a percent is critical to them. But they are also aware that tomorrow will be no better unless they constantly try to develop new processes, enter new markets, and offer superior—even unique—products. As one senior German executive phrased it recently, "We look at rates of return, too, but only after we ask 'Is it a good product?' [9]"

Creating economic value

Americans traveling in Europe and Asia soon learn they must often deal with criticism of our country. Being forced to respond to such criticism can be healthy, for it requires rethinking some basic issues of principle and practice.

We have much to be proud about and little to be ashamed of relative to most other countries. But sometimes the criticism of others is uncomfortably close to the mark. The comments of our overseas competitors on American business practices contain enough truth to require our thoughtful consideration. What is behind the decline in competitiveness of U.S. business? Why do U.S. companies have such apparent difficulties competing with foreign producers of established products, many of which originated in the United States?

For example, Japanese televisions dominate some market segments, even though many U.S. producers now enjoy the same low labor cost advantages of offshore production. The German machine tool and automotive producers continue their inroads into U.S. domestic markets, even though their labor rates are now higher than those in the United States and the famed German worker in German factories is almost as likely to be Turkish or Italian as German.

The responsibility for these problems may rest in part on government policies that either overconstrain or undersupport U.S. producers. But if our foreign critics are correct, the long-term solution to America's problems may not be correctable simply by changing our government's tax laws, monetary policies, and regulatory practices. It will also require some fundamental changes in management attitudes and practices.

It would be an oversimplification to assert that the only reason for the decline in competitiveness of U.S. companies is that our managers devote too much attention and energy to using existing resources more efficiently. It would also oversimplify the issue, although possibly to a lesser extent, to say that it is due purely and simply to their tendency to neglect technology as a competitive weapon.

Companies cannot become more innovative simply by increasing R&D investments or by conducting more basic research. Each of the decisions we have described directly affects several functional areas of management, and major conflicts can only be reconciled at senior executive levels. The benefits favoring the more innovative, aggressive option in each case depend more on intangible factors than do their efficiency-oriented alternatives.

Senior managers who are less informed about their industry and its confederation of parts suppliers, equipment suppliers, workers, and customers or who have less time to consider the long-term implications of their interactions are likely to exhibit a noninnovative bias in their choices. Tight financial controls with a short-term emphasis will also bias choices toward the less innovative, less technologically aggressive alternatives.

The key to long-term success—even survival—in business is what it has always been: to invest, to innovate, to lead, to create value where none existed before. Such determination, such striving to excel, requires leaders—not *just* controllers, market analysts, and portfolio managers. In our preoccupation with the braking systems and exterior trim, we may have neglected the drive trains of our corporations.

References

[1] R. Suzuki, ''Worldwide Expansion of U.S. Exports—A Japanese View,'' *Sloan Management Review*, p. 1, Spring 1979.

[2] *Business Week*, p. 57, Feb. 16, 1976.

[3] B. G. Malkiel, ''Productivity—The Problem Behind the Headlines,'' *Harvard Business Review*, p. 81, May–June 1979.

[4] R. Bennett and R. Cooper, ''Beyond the Marketing Concept,'' *Business Horizons*, p. 76, June 1979.

[5] J. H. Davidson, ''Why Most New Consumer Brands Fail,'' *Harvard Business Review*, p. 117, March–April 1976.

[6] *Business Week*, p. 57, Feb. 16, 1976.

[7] E. von Hippel, ''The Dominant Role of Users in the Scientific Instrument Innovation Process,'' MIT Sloan School of Management Working Paper 75-764, Jan. 1975.

[8] *Dun's Review*, p. 39, July 1978.

[9] *Business Week*, p. 76, Mar. 3, 1980.

15/SUCCESSFUL CORPORATE R&D

Roland W. Schmitt

The top managers of every diversified corporation must decide whether to leave the technical future of each business entirely in its own hands or to do some R&D at the corporate level. The first choice risks too much competition with the short-term demand for profits; the second, too little linkage with the needs of individual businesses. How are managers to make this decision, and—equally important—how are they to evaluate its results?

Experience in managing technology teaches us that the success of corporate level R&D depends on

- A sufficiently high proportion of corporate funding to permit the centralized laboratory to define its own programs
- Close linkages with the plans, strategies, and programs of individual businesses as a basis for its own programs
- A thorough understanding of corporate goals and strategies to guide a balance of its programs among various businesses
- The need to hire the most outstanding people available and to link their compensation and advancement to technical performance, not to some later move into management.

The are several preeminent reasons for doing corporate R&D in the first place: synergy, interdisciplinary focus, and lead time. Some technologies may, for example, be relevant to several businesses. At General Electric, solid-state power technology—the control of electrical power for such diverse applications as industrial drives, lamp ballasts, power supplies, and appliance controls—has applications in semiconductor components, motors, appliances, and automation systems. Corporate research also has value in building interdisciplinary teams to exploit radically new technical approaches. For development into a successful product, the CT scan method of X-ray imaging required interdisciplinary skills and knowledge that were not found together in GE's traditional X-ray business. But they could be brought together at the corporate laboratory.

In addition, centralized R&D helps shorten the lead time for introducing new products and processes because this kind of lab can and should work in fields associated with, but not yet targeted on, specific business objectives. When a target becomes clear, as has happened at GE time and again in such areas as engineering plastics and solid-state power, the talent and ability are in place to move quickly. Indeed, if a company's strategy is to diversify into closely related businesses for which there is appropriate competence in-house, a corporate laboratory tied closely to existing businesses but looking beyond current horizons is essential.

With so much agreement on the reasons and criteria for corporate R&D, we might reasonably expect a further consensus on the best measures for judging both the nature and degree of its contribution in different companies. In practice, things are not so clear-cut.

A difference in approach

Compare, for a moment, the activities of Bell Laboratories with those of GE's Research and Development Center. Without question, Bell Labs has fostered much Nobel Prize quality work—on transistors, lasers, information science, solid-state electronics, and the like—and has had a profound effect on modern technology and industry. Its parent company has captured only a fraction of the vast array of industrial benefits of these advances, although it has been able to earn regulator-approved levels of return on much of its R&D costs.

GE's Research and Development Center also has a distinguished history. The list of innovations from which it has built successful business extends from ductile tungsten and high volume X-ray tubes through silicones and man-made diamonds to computerized tomography for medical applications. In Nobel Prizes and similar honors, Bell Labs is undeniably the leader, while GE's Research and Development Center can claim a high level of productivity in commercial output, producing five times as many patents per scientist and engineer as Bell Labs has.

How can we best understand these differences in approach?

Generic vs. targeted

Some observers view centralized R&D as trying to strike the optimum balance among basic research, applied research, and development. I do not think that these are operationally useful categories. For me, the key is whether a corporate laboratory is working at the forefront of technical areas centrally important to the parent company. Is it producing results of near-term value and laying the groundwork for future advances?

In many areas, it is entirely possible to stay at the forefront of technology by working on targeted developments alone. With robotics, for example, successful design and application require no fundamental new knowledge. Yes, the physical principles underlying such applications as welding or assembly need to be understood. Yes, it takes imaginative engineering to integrate robots with optical and tactile recognition systems, advanced controls, and CAD/CAM-based design and simulation. But these needs can all be met within programs directly addressed to robot applications and products.

So, too, in part, with expert systems and artificial intelligence. Here there is considerable scope for new knowledge about inference mechanisms, knowledge structures, and the best languages in which to express them. But, again, a lab can explore these issues in the context of specific applications. Just think about the organizations carrying out this work: medical schools; financial institutions; companies involved with natural resource location, such as Schlumberger; computer builders, such as Digital Equipment; and equipment builders, such as GE with its expert system for locomotive maintenance.

To be sure, targeted research demands challenging goals that push the limits of knowledge. For technologies like the micron and submicron integrated circuit, however, targeted programs are not enough. When the rate of progress is especially rapid (and new discoveries and inventions are common), it is not possible to ensure forefront competence by concentrating on tightly focused programs alone. In these cases, it is prudent to carry on untargeted or generic research in areas of continuing pertinence to the business. The supreme example of such an area is biotechnology, where most of the ultimate applications are still highly speculative. Today, the status of biotechnology is much like that of solid-state and semiconductor physics 30 years ago when Bell Labs, GE, and a host of other organizations mounted significant programs of generic research. Had they not, the targeted VLSI research and development of the present would not be possible.

Generic research is not, however, synonymous with a license to work on anything that happens to be interesting and fruitful. Many fascinating and very productive areas of science, such as high-energy physics, astronomy, and cosmology, are completely outside the purview of industrial R&D, generic or not. This is as it should be, since it is impossible to connect these areas of research to any viable business, now or in the foreseeable future. As a rule, U.S. industry undertakes only that research for which applications can be envisaged, even if these applications are ill-defined and remote. Lasers, for example, largely resulted from generic industrial R&D: their applications, which are now so important, took decades to fully emerge.

Bell Labs is justly famous for its generic research; GE's corporate lab, along with the corporate labs of many Japanese companies, for its targeted work. The Japanese also tend to pursue alternative approaches to their R&D targets, using redundancy to increase the probability of success. Although U.S. companies sometimes follow suit, the Japanese carry the practice much further along the development cycle. The U.S. company may be more efficient, but the Japanese will be more confident of achieving the goal. The history of consumer electronic products like videocassette recorders illustrates this process very well.

There is a useful distinction to make between this emphasis on redundancy and the approach of Western entrepreneurs who typically eschew both generic R&D and redundancy and concentrate instead on the final stages of targeted development. With a rich lode to mine in the generic, untargeted, and as yet underexploited research of major companies or universities, these smaller companies are almost entirely end-use focused.

Consequently, the larger companies that do generic research always run the risk of seeing it exploited by entrepreneurial organizations before it can make a real contribution to the businesses of the parent. Research in catalysis has often fallen into this category; today, a prime example is computer language research. The software for Apple's MacIntosh computer was derived in large part from basic work done at Xerox's Palo Alto Research Center.

Market-driven vs. technology-driven

A second key issue that observers often raise about corporate R&D is whether it should be guided by knowledge of market requirements or by perceived capabilities of technology. Correctly anticipating future markets is every bit as uncertain as attaining technological goals. Even existing markets, whose size and rate of growth are well known, may respond in unexpected ways to new technology. Latent markets, which do not exist but might develop in response to a new product, pose even greater risks—and opportunities. In materials-based industries, several innovative companies—Corning, 3M, and GE's plastics, silicone, and industrial diamond businesses among them—have been quite successful in creating and exploiting such markets through a combination of generic and targeted research.

To be effective, industrial R&D must, by definition, be linked with successful marketing. The real issue, then, is not the choice between a market focus and a technology focus but the proper mode of coordination between them.

Marketing experts should be poised over the new ideas that emerge from generic research, but they should also be involved in the identification of targeted programs. These experts should not, however, give blind allegiance to the latest analytical techniques or the dogma of marketing supremacy. Rather, they should have the temperament of a research experimentalist, putting forth hypotheses about the market and devising economical and efficient market experiments. The Japanese have ably practiced this experimental approach, especially in audio and video electronic products and computer memories. They have taken chances by introducing still questionable product innovations, listening to customers' responses, and tailoring the product accordingly. In the United States, it is far more common to conduct one market research study after another, with none giving sufficiently clear answers to support an unequivocal management commitment to go to market.

In recent years, formal business planning has stressed the risk-weighted, discounted rate of return on R&D projects. The professional literature bulges with methodologies for R&D project evaluation that focus on estimates of potential markets, project costs, and probabilities of success. For slow-growing and protected industries, these methods have been successful. For fast-moving and worldwide industries, they have been a disaster. The Japanese and Western entrepreneurs have outflanked and overwhelmed companies relying on such hands-off analytical models. It is inconceivable that today's successful software and computer peripheral producers—or tomorrow's biotechnology companies—started with a careful analysis of ROIs on their research projects.

What I have said about markets is somewhat less true about in-house needs for process improvement. After all, it is much easier to spot opportunities for cost reduction, improved testing, and quality control than it is to spot new markets. Process advances, however, often require highly sophisticated research and development involving a multiplicity of disciplines. The massive retooling and automation that companies such as General Motors have accomplished would have been difficult without a massive base of internal R&D.

Two kinds of success

One obvious difference between corporate R&D at Bell Labs and at GE is that Bell Labs is at least ten times larger. I do not believe, however, that size bears any direct relation to success, with the single caveat that for each technical area there does exist a minimum "critical mass" of expertise. In view of Bell's record of achievement, it is hard to have too large a laboratory; still, there are many examples of small but highly successful industrial laboratories—in specialty chemicals, in computer peripherals, in semiconductors, and elsewhere.

The main differences, then, are not related to size but to operating philosophy: for Bell Labs, an emphasis on generic programs and system efficiency in regulated markets; for GE, an emphasis on targeted programs and close coordination with highly competitive markets. Are both organizations equally successful? Obviously, the answer depends on the parameters used to measure success.

For contributions to science and technology generally, Bell stands out, although it is far from clear what the future holds for the labs in the newly deregulated universe of AT&T. One of the great ironies of modern times may be the destruction of the historic role of Bell Labs at the very moment we are most concerned about U.S. competitiveness and the revitalization of U.S. industry, to which that lab has contributed so much. For contributions to its parent company, I would say that the General Electric R&D Center comes out very well. The profits earned today by businesses directly based on corporate R&D run some four to five times the annual cost of the GE lab.

The profits of these businesses over and above the company average run almost twice the annual cost of the lab. And this measurable output is only a fraction of the total; most output is in the form of new and improved products or processes that are incorporated into existing businesses and whose effect is thus much more difficult to measure directly.

To be sure, both organizations have common elements underlying their success: the quality of people, first-rate shop and maintenance operations, and excellent systems of program management and communication, both upward and downward. Corporate R&D can fail for many reasons—not enough good people, domination by external contracts or customer pressures, lack of top management support. When it flourishes, however, the prime reason is less the choice of operating philosophy itself than the organization's knowledge of—and confidence in—that philosophy.

Every industrial laboratory must know its goal, believe in it, and organize for it. All too often, there is no agreement between corporate R&D and top corporate management on what the lab's mission and purpose are. In many companies, technical managers seek to promote first-rate science while the corporate office initiates study after study to find out why no new businesses have resulted from centralized R&D. Or the labs make significant contributions to their companies while executives fret that the labs do not earn kudos for the corporation among security analysts, customers, prospective employees, and so on.

This is not, of course, an either-or situation: centralized industrial R&D organizations do not have to be replicas of either Bell Labs or GE. It is quite possible to have highly successful laboratories in which some areas of work are targeted and some generic. In fact, neither Bell Labs nor GE's R&D Center is a monolithic example of generic or targeted research. Bell Labs has made unique contributions to its parent's businesses in communication systems design and in microwave technology; General Electric sponsors generic research in areas as diverse as biophysics and coal science. It, too, has had its share of Nobel laureates.

Further, the balance of targeted and generic work depends on how centralized the company's involvement in research and development is. Some major corporations like RCA, Kodak, and Sohio concentrate R&D heavily in a single laboratory; others like IBM, Monsanto, Xerox, and GE have several. Success is possible in either mode. A single laboratory must have a mix of generic and targeted research. In a multilaboratory company, some laboratories can be more generic, some more targeted. However, researchers must know which is which and must understand what their roles are in the overall corporate strategy.

The success of a corporate R&D program becomes visible only in the light of its mission and purpose. If the choice is to adopt a generic, loosely market-coupled approach, then organization requires a strong discipline orientation and close attention to the number and excellence of contributions to the technical literature. If the choice is to adopt a targeted, tightly market-coupled approach, then organization needs a project orientation and must link its rewards to ultimate business success. Failure inevitably comes from trying to organize, appraise, and reward according to one approach while expecting results typical of the other.

R&D personnel must understand and agree with the purposes inherent in their jobs if they are to be both productive and high spirited. People working in generic programs have to recognize that they are being asked to produce new knowledge and will be appraised accordingly. People in targeted programs have to recognize that they cannot take irrelevant detours into interesting science merely out of their own curiosity. Both must accept the challenge to work in different ways at the forefront of technology. Although they take different forms, both kinds of success are essential to the long-term health of American industry.

16/A REVIEW OF PROGRESS AT MCC

Mark A. Fischetti

The drums of publicity reverberated loudly when the Microelectronics & Computer Technology Corp. was formed—but not for long. After being set up in 1983 by 10 otherwise competing U.S. computer and semiconductor companies to upstage Japanese computer advances, MCC remained silent about its work. After declaring its goal—making the founding shareholders able to introduce "fifth-generation" computer technology before Japan did—it faded from public view.

Where is MCC today? Not only alive and kicking as a cooperative research effort in Austin, TX, but with doubled membership and a redefined goal. No longer is the company chasing the elusive computer that was to revolutionize the industry by incorporating artificial intelligence and making decisions based on reasoning. Along with most of the rest of the U.S. electronics industry, MCC's goal now is to get the same effect by developing seven areas of attainable computer technology (Fig. 16-1) and linking the results.

The areas of concentration are:

- New semiconductor packaging to increase the density and decrease the size of systems containing ICs.
- More efficient software design by large software teams
- More efficient IC design through computer-aided design
- Parallel processing machines for faster computation
- Artificial intelligence and knowledge-based systems for computers that can "reason"
- Large, complex databases to store the mass of information required by knowledge-based systems
- Better interfaces between computers and people, so users can work more effectively.

The 10 original shareholders were jarred into action by Japan's announcement of a nationwide, government funded, crash research

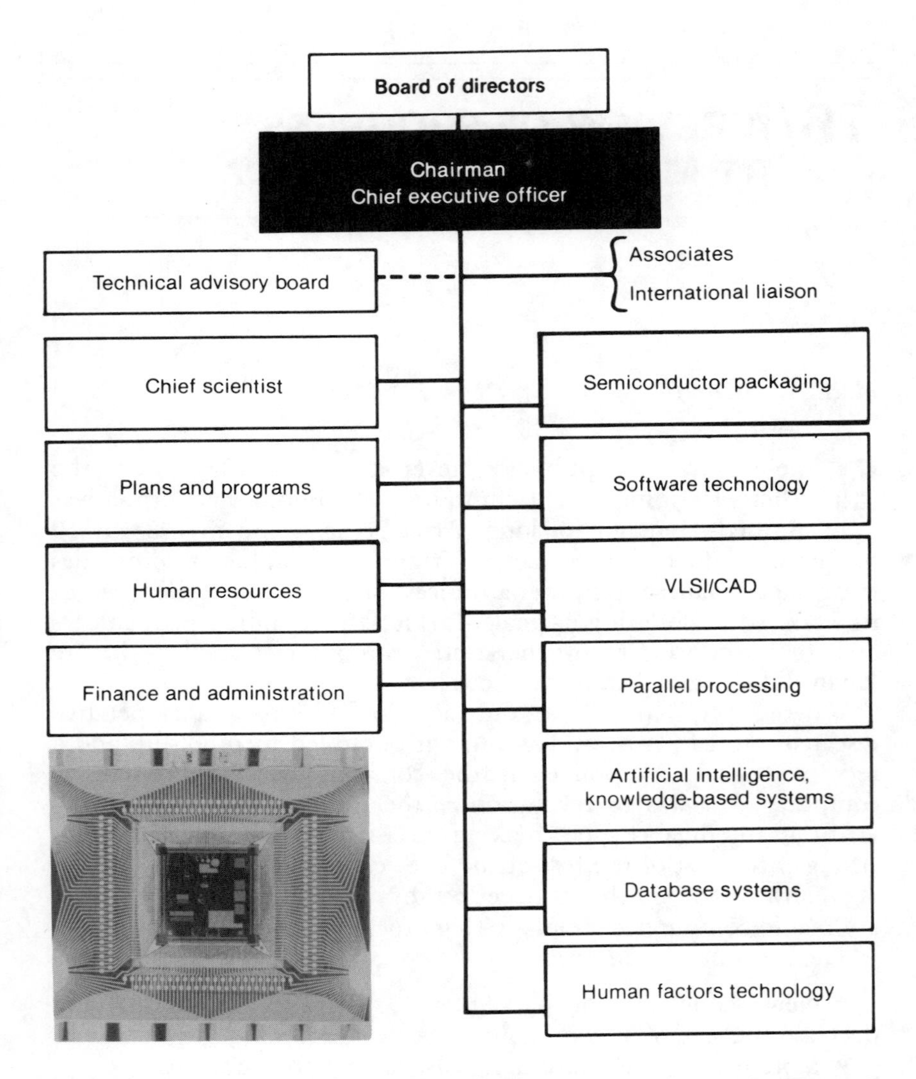

Fig. 16-1: *MCC is governed by a board of directors—one representative from each of its shareholders and Bob Inman, the chairman and chief executive officer. Advice on research strategy and new proposals is provided by a technical advisory board of senior executives from the shareholders. Developments outside the United States are analyzed by the international liaison office.*

Inside MCC, John Pinkston and Palle Smidt oversee technical progress and coordinate the seven research programs. Companies that cannot afford or choose not to join MCC full-time can share in research symposiums and receive nonproprietary reports by paying fees for an associates program.

An early result from the packaging program is the tape-automated-bonding test device shown. An IC placed at the center is bonded to the leads on the two-layer film made by 3M; the chip and leads are then cut away and bonded to a circuit board. This device has 172 leads centered 7 mils apart. On a circuit board it would cover 0.36 square inch, whereas a conventional dual-in-line IC package with only 40 leads would cover 1.2 square inches.

program to develop fifth-generation computers before the United States, positioning Japan to dominate the world computer and semiconductor markets in the 1990s.

MCC was formed quickly, in an atmosphere of crisis. It began operations on Sept. 1, 1983, as a commercial company, performing proprietary research. Plans were to assemble a top staff by borrowing researchers from the member companies. When a research program was completed, staff members on loan were to return to their companies, taking the technology along.

A flurry of newspaper and magazine reports described MCC's formation. Since then, few details have been publicized. One researcher engaged in efforts similar to MCC's summed up the knowledge most outsiders have about the company's work: "Nobody knows what's going on over there."

MCC agreed to let *Spectrum's* staff visit its headquarters for a broad briefing and for interviews with researchers and executives who described MCC's strategy and how it functions.

Japan's progress not a factor

First of all, "beating Japan" is not MCC's goal. "The Japanese fifth-generation project clearly acted as the catalyst for our shareholders," said chief scientist John Pinkston, who was formerly deputy chief of the U.S. National Security Agency's research group. "But having created MCC, we haven't shaped up to be particularly anti-Japanese. In our planning process, it hasn't been: 'What should we do because the Japanese are doing it?' It's been: 'We should do this because it would lead to performance better than anyone else's.' If the Japanese were to fold up their tents and go away tomorrow, MCC would still be a good thing for the shareholders."

Nonetheless, the shareholders, who must concern themselves with international competition, are breathing a little easier with MCC on their side. "We don't have the same fear," commented G. Michael Schumacher, vice president of software and technology at Control Data Corp. and that company's representative on MCC's advisory board. "Japan, however, should not be taken lightly," he warned.

MCC's work, Pinkston explained, will not culminate in a specific fifth-generation computer. It may advance computer technology to the point of prototypes, he said, but conversion into marketable products rests with the shareholders.

One of the first problems the cooperative think tank encountered was assembling the research staff (Fig. 16-2). The idea that MCC would borrow most of its researchers from shareholders simply did

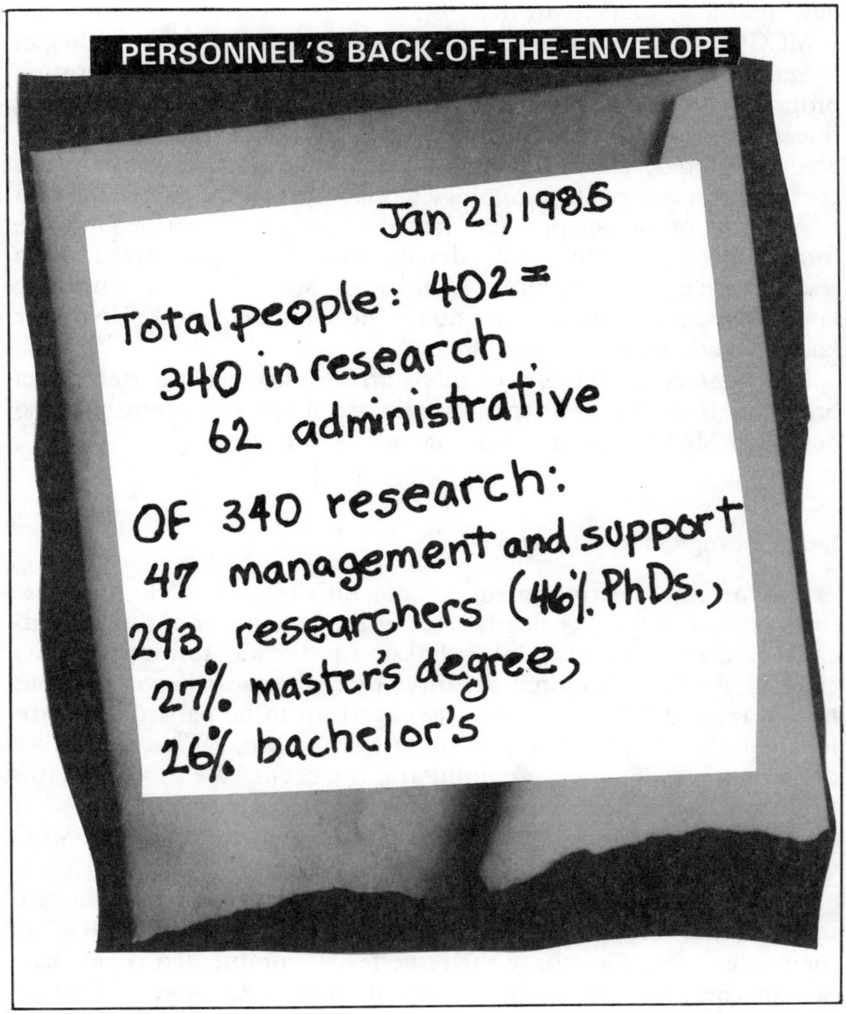

PERSONNEL'S BACK-OF-THE-ENVELOPE

Jan 21, 1986

Total people: 402 =
340 in research
62 administrative

OF 340 research:
47 management and support
293 researchers (46% PhDs.,
27% master's degree,
26% bachelor's

Fig. 16-2: *Gathering MCC's large top-flight research staff took more than the first year of management's time, the reason that MCC "did not get off to an auspicious start," said the company's vice president for human resources, George Black. He added that the wait was worth it. The 402 employees now at MCC will be joined by 60 more within six months, Black said, as MCC leaves the planning stage and rolls up its sleeves to develop hardware and software.*

not pan out. Some shareholder companies were reluctant to let their top talent leave. At present only one-third of MCC's researchers are from member companies.

MCC leaders also soon realized that shareholders could not wait for an MCC research program to end before their employees would return with the fruits of their labors—by then it would already be too late. Bob Inman, former deputy director of the Central Intelligence Agency and now chairman and chief executive officer of MCC, said the unforeseen need to recruit so much outside talent took up most of his time and the company's energy for the first 18 months. It put MCC's initial progress some six months behind schedule, he said. But the realization that the research team wouldn't include as much talent from the shareholders focused attention earlier on transferring MCC technology to the member companies.

The original MCC format for the research itself was also changed. At the beginning, shareholders joined one or more of four distinct research programs: IC packaging, software technology, computer-aided design (CAD), and advanced computer architectures. These programs were to last 6, 7, 8, and 10 years, respectively.

But since the long-term goal is no longer a single fifth-generation computer, it is now assumed that, although the original milestone deadlines still apply, the programs will continue indefinitely. "The numbers 6, 7, 8, 10 had to be there in the first plans," Pinkston said. "People were unwilling to sign up on an open-ended basis."

"The compartmental setup was a vehicle to get MCC off the ground," added Palle Smidt, who as senior vice president for plans and programs is the key liaison between MCC and the shareholders.

Smidt, previously the vice president of business strategy for Sperry Computer Systems, stressed, however, that specific "deliverables"—the MCC term for prototypes, algorithms, designs, or any other tangibles to be produced—would be made available according to the original schedules.

The original areas of research have long since expanded also. The program in advanced computer architectures (ACA), for example, has been split into four full programs: parallel processing, artificial intelligence and knowledge-based systems (AI-KBS), databases, and human factors technology. (The four programs remain linked; shareholders who join ACA are enrolled in all four areas.) The four computer programs plus packaging, software, and CAD give MCC its seven programs.

Seven is not considered a magic number, however. The door is open to more research. Indeed, last January the board of directors approved a new "integration group" composed of up to nine researchers. It is to function as the fifth element of the computer architectures group. Its staff will monitor the results from the four computer programs, acquire data from trends outside MCC, and begin to design the kind of computer system or systems that might evolve from architecture programs, Pinkston said. Eventually, he added, the group is to be responsible for integrating all five computer architecture programs.

"I don't believe we're talking about creating *one* architecture," Pinkston said. "I think there are going to be a bunch of them. And the systems may evolve toward being specific to classes of applications."

This is in line with MCC's ultimate goal, articulated by Inman: "In the 1990s we aren't talking about a computer industry, or a semiconductor industry, or a telecommunications industry. We're talking about an information handling industry. MCC ought to evolve in programs that move toward a systems approach to information handling."

As evidence of this trend, Inman offered a profile of the types of companies joining MCC. The founders were largely computer manufacturers and semiconductor makers. Those that have come aboard since are much broader-based, like Bell Communications Research, the Boeing Co. Inc., and 3M.

Technology transfer is the key

The success of shareholders in applying technology, and the success of MCC in preparing them for the 1990s, depend on one key element: technology transfer.

Transfer to shareholders would best take place by integrating research results among the programs. Such sharing was originally forbidden, however, because different sets of shareholders support different programs, and proprietary interests would be jeopardized by swapping information.

Yet each shareholder knows it can get more bang for its buck by integrating research. Improvements in designing workstations that come out of the human factors program could greatly aid VLSI-CAD progress, for example. Therefore, the board of directors has approved means for concluding formal agreements between shareholders in related programs. As actual research results are transferred, the agreements are turned into licensing arrangements.

Such integration is likely to accelerate as research approaches the prototype stage and as the 1990s draw closer. At that time, another type of transfer will take on added emphasis—transfer of program results into shareholder development programs.

"It's not enough just to do something fancy," Smidt emphasized. "If we do not transfer technology effectively, we will still have not done the job."

Irwin Dorros, executive vice president of technical services at Bell Communications Research and that company's representative on MCC's board of directors, concurred. "We don't take for granted their success," he said, adding that the ultimate measure is "how well they transfer technology to the shareholders so it influences their products."

As for "interim deliverables," there have been "successes we had not planned," Inman said. "Dividends of 1985 occurred where we had put together a unique equipment configuration or an algorithm as ways to attack a problem, and the shareholders said, Hey, we'd like to use those in our research at home."

For example, engineers in the packaging program built a new machine to automatically process tape-automated-bonding chips, speeding their production. Researchers in the database program developed a new computer language called LDL for use in database machines.

Another problem in the transfer area is that MCC is prevented from setting standards by U.S. antitrust laws. Shareholders are commercial competitors and were they to set standards through MCC, they could be found guilty of collusion.

"But we can provide data on which a standards decision could be based," Pinkston said, "although we cannot be part of that decision process." If MCC does provide such data, he said, shareholders would then create standards through the conventional standards-setting forums.

Pinkston noted that the problem with standards also occurs in reverse—technology being developed by MCC does not necessarily conform to standards used by shareholders. Computer code being compiled in the CAD program, for example, is being written in the Lisp workstation language, he said, "but the number of shareholders running CAD systems now on Lisp is zero."

Competing with other co-ops

One of MCC's strong points is its relatively simple hierarchy: Inman, the executive committee (Inman, Pinkston, and Smidt), the program directors, and the researchers. This keeps bureaucracy down and creates the climate for rapid decisions, Pinkston said. "It can be terribly frustrating for a researcher to have an idea he knows how to pursue and then to have to sit and wait," he observed.

MCC also gives its directors wide range for expenditures on equipment or other supports, Pinkston noted. "Program directors can approve expenditures of up to $100,000," he said, and use their budgets as they see fit; they are not locked into spending so much for people and so much for capital expenses.

Although MCC takes extra care to ensure smooth relations within its ranks, its stand on relations with other research consortia is much different. MCC was started at a time when much national attention was focused on cooperative research. Companies in the semiconductor and computer industries were finding they simply could not afford the growing cost of basic research, so they began to share the expense and the risk by cooperating in research with other companies. Within the period of MCC's creation, other cooperative research ventures were formed—the Semiconductor Research Corp. in Research Triangle Park, NC; the Microelectronics Center of North Carolina, also in Research Triangle Park; and the Microelectronics and Information Sciences Center in Minneapolis, MN, and others.

These groups operate as consortia, channeling funds from corporate and government sponsors to laboratories and universities engaged in research. When formed, they hoped that by sharing results they could bolster U.S. industry while upgrading research at universities. Another intent was to coordinate their work to eliminate duplication. They had hoped, too, that MCC would join them. But it did not, and it will not.

"MCC does not try to fit into a collection of research activities," Smidt said. "Our focus is to understand what can derive the potentially biggest revenue opportunities and profit for our shareholders, irrespective of what the rest of the world does."

MCC, he stressed, is a commercial research venture, not a national resource. "National objectives and company business objectives are not necessarily the same," he said. "People form businesses to increase their wealth. If they in the process also meet national objectives, it's wonderful."

Is MCC then competing with the other consortia? "De facto, yes," Smidt said. However, in terms of actual research, MCC's work does not seem to be duplicating that of other cooperatives, Pinkston said, except possibly some of the packaging and design work at Semiconductor Research Corp.

Software researchers also question whether there is overlap between MCC's software program and the research at both the Software Engineering Institute of Carnegie-Mellon University in Pittsburgh and the new Software Productivity Consortium in Washington, DC. Laszlo Belady, director of MCC's software program, said each group aimed not to duplicate research elsewhere but noted that they were competing for the same researchers.

Smidt added that MCC will certainly air its views on persisting problems in technology research and discuss its own research, "but we will not inform the world as to what our solutions are. That's what our shareholders are paying for."

Can MCC fail?

Operating independently of a larger national research effort, making no products itself, and having extended the length of its programs, MCC is essentially free of obstacles that sometimes hamper research. Can it still fail?

"Absolutely," Smidt said. Although he maintained that this is not likely, he added: "MCC can fail by doing research that is irrelevant to the goals its shareholders are pursuing. And it can fail by being late—timeliness is very important. MCC can also fail due to the normal cause of failure—ineffective, inappropriate, and unskilled management." Finally, Smidt said, "if we did all the right things in the world, at the right time, we could still fail if the shareholders do not capitalize on the excellent output."

Smidt and Inman also noted that it was possible for programs or segments of programs to come up empty-handed, but added that even this was of value to shareholders by preventing the waste of their resources.

That attitude underlies the very nature of MCC's high-risk research, and all involved understand this, Inman said. "We weren't created to proceed cautiously," he said, because high risk portends high payoff. "There will be failures. The key is when to cut them off and decide they're just not going to pay."

Technology: from lab to shareholder

As a loner in the competitive world of corporate research, MCC has naturally kept the details of its work secret. Last January, however, the technical program vice presidents gave *Spectrum* a rough outline of the structure and purpose of their programs, and cited evidence of progress so far (Fig. 16-3). Much of the research was described by Pinkston as exploratory; more basic research in being done in artificial intelligence and advanced development in the packaging program. Here, broadly, is how the seven programs are shaping up.

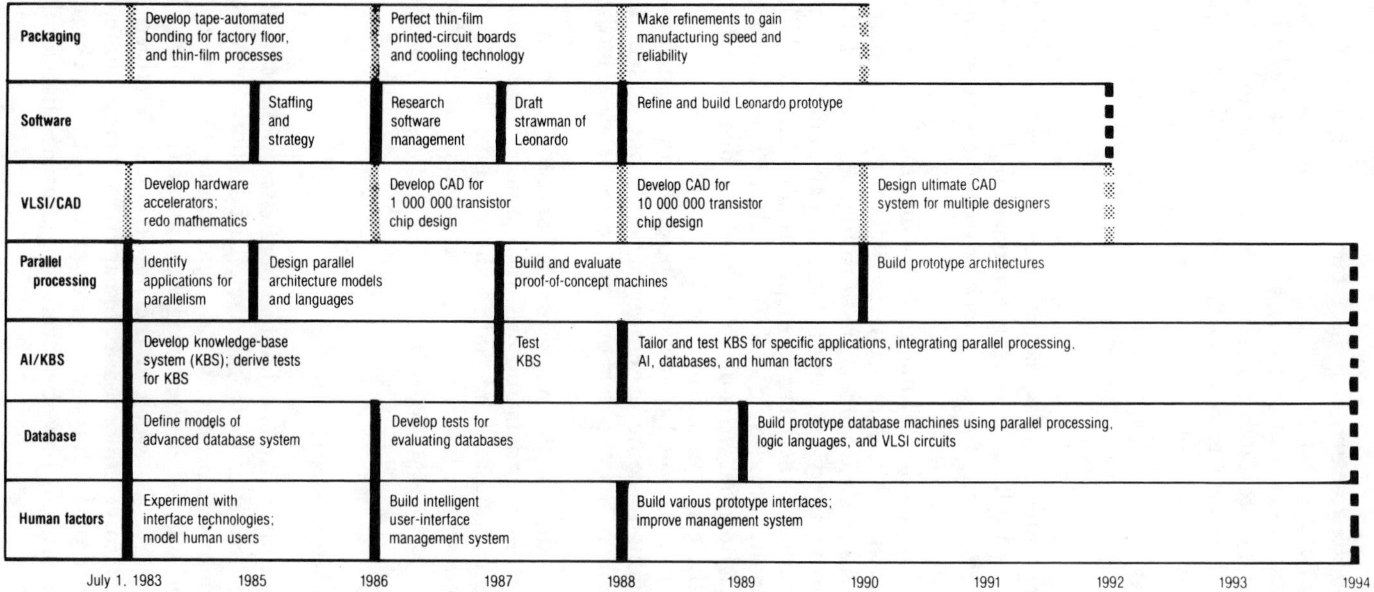

Packaging	Develop tape-automated bonding for factory floor, and thin-film processes	Perfect thin-film printed-circuit boards and cooling technology	Make refinements to gain manufacturing speed and reliability	
Software	Staffing and strategy	Research software management	Draft strawman of Leonardo	Refine and build Leonardo prototype
VLSI/CAD	Develop hardware accelerators; redo mathematics	Develop CAD for 1 000 000 transistor chip design	Develop CAD for 10 000 000 transistor chip design	Design ultimate CAD system for multiple designers
Parallel processing	Identify applications for parallelism	Design parallel architecture models and languages	Build and evaluate proof-of-concept machines	Build prototype architectures
AI/KBS	Develop knowledge-base system (KBS); derive tests for KBS	Test KBS	Tailor and test KBS for specific applications, integrating parallel processing, AI, databases, and human factors	
Database	Define models of advanced database system	Develop tests for evaluating databases	Build prototype database machines using parallel processing, logic languages, and VLSI circuits	
Human factors	Experiment with interface technologies; model human users	Build intelligent user-interface management system	Build various prototype interfaces; improve management system	

July 1, 1983 1985 1986 1987 1988 1989 1990 1991 1992 1993 1994

Fig. 16-3: *Six of MCC's technical programs formally began on July 1, 1983, and the software program started one year later. Each program was to last a specified number of years. While the original objectives must still be reached by the program deadline, MCC shareholders and leaders have already agreed to extend the programs themselves. The timeline steps are not strictly serial; work for even the last steps has in most cases already begun. The divisions highlight both the major concentration of work in a given time period and the accomplishments needed to keep progress on schedule. (MCC officials have not released detailed schedules; the program highlights on the timeline were compiled by* Spectrum *based on MCC interviews.)*

Semiconductor packaging

New and improved chip and circuit board packaging is the basis of MCC's shortest technical program, originally scheduled to last six years. It is also the most aggressive in terms of "deliverables," having already produced several tangible benefits.

Much of the work in integrated circuit packaging is focused on turning the experimental tape-automated-bonding (TAB) method into a robust production-floor process.

"There have always been sexy ideas for packaging chips," Barry Whalen, formerly of TRW, said. "The problem is turning those ideas into equipment that can be used on the manufacturing floor." TAB has been used for a few years by a handful of U.S. and Japanese semiconductor companies to package ICs with a small number of leads; MCC is looking to foster widespread TAB use for high-lead-count chips.

TAB embodies several advances in chip packaging. With most techniques in use today, a chip is bonded within a closed, hard plastic package. Package leads are soldered onto printed-circuit boards. With TAB, the chip is bonded to the surface of a tape closely resembling 35-mm camera film (Fig. 16-1). Tiny connections are formed between special pads on the chip and leads on the tape by depositing metal.

The chip and its leads are then sliced from the tape and bonded directly to a circuit board.

The net result: chips can be placed much closer together on boards. Some 35 to 40 percent of the board area is covered by TAB chips made at MCC, Whalen said, compared with an average of 1 to 5 percent for printed-circuit boards containing chips in conventional and surface-mounted packages.

The new approach also permits the manufacture of much denser circuits, Whalen noted, because the leads can be placed much closer together. In most production chips, the external leads are 100 mils apart, and a few chips are being made with 50-mil spacing. Chip leads on the MCC TAB line are 7 to 12 mils apart, and next year they will be down to 5 mils, Whalen said.

Chips with 328 leads have been fabricated at the packaging laboratory in the same space needed to provide only 48 or 64 conventional leads, noted Brad Nelson, an MCC researcher.

Circuit boards stand to improve also. By depositing chip-connecting circuitry lines directly onto a board with sputtering machines, Nelson said, MCC has already made adjacent lines 10 micrometers wide with only 10 micrometers of space between lines. This compares, he said, with 75-micrometer-wide lines and 75-micrometer spaces on the best printed-circuit boards.

The payoff with TAB is that denser chips with many more leads can be placed much closer together, on boards with denser interchip circuitry. The ultimate goal, Whalen said, "is to take the circuitry of a supercomputer and put it in a portable system like a workstation."

This assumes, of course, advances in cooling technology; supercomputers like the Cray II are cooled with tanks and tubes of liquid fluorocarbon—hardly the technology for a desktop computer. Whalen said he had begun to "squash supercomputer cooling technology down" and might end up with a system that uses active water cooling.

A key element in making TAB a production-line technology is inspection and testing of chips to ensure reliability. Researchers are investigating noncontact testing, using electron beams, lasers, ultrasonics, and X-rays. These do not compromise reliability as much as methods that physically manipulate the chip.

About 10 percent of the packaging program is dedicated to improving conventional and surface-mounted chip packaging. One project, for example, focuses on reducing the space between leads on pin-grid arrays to less than 50 mils.

One early deliverable is a machine to handle a continuous tape of TAB chips, spooling them through automated test equipment like film through a projector. The machine was needed but simply "didn't exist before," Whalen said.

Though not explicitly stated, MCC is in a race with Japan for better packaging. In Japan, Sharp Electronics Corp. is producing thin-film circuits bonded to tape for use in credit card-style calculators and other products, and NEC Corp.'s latest supercomputer, the SX2, has many of its chips bonded on tape. Being a supercomputer, the SX2 was assembled to a great extent in laboratories, but mass production is believed to be forthcoming.

Software

Since MCC's seven-year software program is one year younger than the others, the software researchers have so far been occupied in defining their challenge and the approaches to it.

"We are not interested in data processing or scientific applications, or how to program for personal computers," said the software program director, Laszlo Belady, formerly with IBM. "We will develop software and a software-producing environment for very large, distributed, concurrent, real-time computing."

Although applications are relatively rare today, these massive software team efforts will become central to computing in the later

1980s and in the 1990s, Belady said. He emphasized that for now his researchers would concentrate on designing software—the "upstream" end of software development—and not on deciding which programming language to use or how to translate specifications into machine code—the "downstream" end.

To increase the productivity of software designers, Belady's group will focus on developing the best computer-aided environment for software design teams, rather than on full automation. This graphics-oriented environment is called Leonardo by the MCC researchers, after Leonardo da Vinci.

The program, Belady said, will culminate in a prototype—perhaps a network of workstations and other tools. The work has been divided among four groups. The design process group is using behavioral science to model how teams of engineers produce software and to define areas of inefficiency. The group will then describe the functions that can be computerized.

The design information group will use this description to develop aids for the software designer, perhaps using expert systems. The design interface group will then seek to improve the way designers input software into a computer and the graphic portrayal of software. Animation, for example, might be used with an expert system to show the effect that a change in the middle of a software program would have on the end of the program. The fourth MCC software group—the design environment group—is taking the lead in building Leonardo. Interim deliverables that have already been generated include an algorithm to solve the so-called N-party interaction problems.

Belady declined to say what software language the developers would use. When the time comes to adopt a language, he said, it will be clear which one is best suited, based on developments outside MCC. He noted that whatever language is chosen, Leonardo could be adapted to it. For example, Belady said, "perhaps all of Leonardo will be translated into Ada," which is the official software language of the U.S. Defense Department.

VLSI-CAD

The first thing to realize about the VLSI-CAD program is that no one is building VLSI circuits; the program is instead developing CAD tools for designing VLSI circuits.

All programming in MCC's CAD program is being done in the Lisp language, which relies heavily on graphics to represent data and circuit blocks. "We are pushing a total graphics approach to design," said the program director, John Hanne, formerly with Osborne Corp., Texas Instruments, Inc. He added that Lisp enables rapid prototyping.

The near-term goal is to develop a CAD system for designing chips with 200,000 devices, but the real payoff will be systems that can handle chips containing 10 million devices—the goal for 1990. Hanne pointed out that these numbers refer to full logic chips, not merely memory ICs.

To accomplish this, Hanne and his staff will configure a system with multiple interactive workstations that communicate with specialized high-throughput computers, or ''accelerators,'' which MCC will build.

All design tasks in the network system will be controlled from Lisp workstations. A key task is to develop a common user interface in which the same function (such as ''move a wire'') will be achieved by the same command in different workstations. Today, most CAD ''systems'' are simply stand-alone workstations linked together in a piecemeal fashion.

CAD is being developed for tasks from process simulation to VLSI layout and system synthesis. Layout programs are geared to the so-called floor-plan approach to design, in which designers pick functional circuit blocks from a library of blocks and concentrate on how to interconnect them. The floor-plan-approach is seen as the fourth generation in design evolution—the first three being gate arrays, standard cells, and logic arrays—and is just beginning to be used in industry. MCC will not pursue gate-array, standard-cell, or printed-circuit board design, although workstations will interface with shareholders' existing design systems for these technologies, Hanne said.

Hanne and his staff have linked 81 Lisp machines to form a hardware system on which to develop software. At the same time researchers are reworking the basic mathematics that generate design algorithms, circuit models, and circuit simulations. Later in the program, Hanne said, knowledge-based systems might be used, for example, to take design-for-test rules and impose them on a circuit design automatically.

To date, Hanne reported, researchers have used the Lisp language to create a ''module editor'' in the programming language ''C'', a computer program that itself allows designers to lay out circuitry graphically. The output is an extension of the Edif language, which consists of lines of mathematical code that drive production processes. The program has 117,000 lines of code and runs on a Unix system. An algorithm has also been developed that solves the so-called reconvergent fanout problem, which has stymied simulation tests of circuit designs.

As work progresses, source code will be transferred to shareholders. MCC will design but not build CAD hardware; hardware implementation will rest with the shareholders, Hanne said.

Parallel processing

Developing computers that process data in parallel rather than serially is both a hardware and software problem, MCC recognizes. Such parallel, or concurrent, machines offer much faster computing speeds.

The few parallel processors that have been built can only execute a particular function and use limited forms of parallelism to solve primarily numeric problems. The underlying principles of parallel processing have not been developed so a computer designer has the option of using it for a range of applications, especially those that use symbolic processing (including expert systems), explained the program director, Stephen Lundstrom, formerly of Stanford University.

An early task for MCC, then, was to understand how to describe computer applications in parallel form, by selecting a number of applications (including Mycin, an expert system that aids medical diagnosis in certain areas, and a VLSI routing algorithm) and determining what concurrency existed or could exist in them.

In addition, a number of programming languages that have potential parallel execution semantics were examined for several test applications. The selected applications were written in a parallel dialect of Lisp (called MultiLisp) and in Prolog. Others were written in SASL (for St. Andrews Static Language), a functional language with graph reduction semantics developed by David Turner at St. Andrews College in England, and in Sisal (for Streams and Iteration in a Single Associated Language), a data-flow-oriented language developed by researchers at Lawrence Livermore National Laboratory, Digital Equipment Corp., the University of Manchester in England, and elsewhere.

Recently researchers have begun developing advanced languages, evaluating various parallel execution models for the languages, and designing physical computer structures to implement those models. Lundstrom said that by the end of 1986 he will initiate the first in a series of proof-of-concept machines, in which different parallel architectures will be tried.

However, MCC will not build parallel computers. Rather, various architectures will be transferred to the program shareholders, which will build parallel computers adapted to specific classes of applications. The goal is to produce high-speed, parallel processor architectures that use symbolic languages and reduce development and execution time by a factor of 10 or 100.

AI-KBS

The program in artificial intelligence and knowledge-based systems (AI-KBS) is perhaps the most adventurous of MCC's efforts. As such, it is the least quantifiable. A primary goal is to develop a computer that can make decisions based on commonsense reasoning as well as facts, solve problems in spite of uncertain or missing facts, and understand human languages.

To date, artificial intelligence has been most visible in so-called expert, or knowledge-based, systems—computer programs that give the same answers to questions that a human expert would. Examples include the Mycin medical program and a military program that identifies safe cross-country pathways for tanks based on recent weather reports and satellite imagery. As advanced as these expert systems are, however, they are limited to solving very clearly defined problems by following the rules dictated by humans—in Mycin's case, physicians.

''What's missing from knowledge-based systems is commonsense knowledge,'' said the AI-KBS program director, Woodrow Bledsoe, formerly of the University of Texas at Austin. In Bledsoe's mind, a medical knowledge-based system, for example, should be able to answer not only: What is the treatment for body temperature of 100°F, swelling glands, and rapid heartbeat? but also: If a hundred-pound box is dropped on a human foot, will there be pain?

A skeleton commonsense knowledge base should be built this year under Doug Lenat at MCC, Bledsoe said. It will include a broad but shallow base of knowledge that humans consider common sense. Included will be knowledge about the knowledge base itself, allowing it to ''know'' when a question is beyond its current understanding.

Computer programs that can understand human, or ''natural,'' languages have also eluded the computer industry for some time. Researchers are investigating the rules that govern how people automatically understand each other.

In order for ''thinking'' machines to be useful they must also produce results in a matter of seconds or minutes, not days. MCC is developing techniques that will dramatically speed up the complex programs that process knowledge bases. Researchers have already developed special computer languages that will allow artificial intelligence techniques to be embedded at the fundamental, or machine-language, level of a computer.

The technology produced will allow MCC shareholders to build various knowledge-based systems, for applications ranging from helping small businesses determine their marketing strategies under different business conditions, to helping defense officials simulate battle scenarios. A test application already underway, Bledsoe said, is a system that will aid designers of IC chips.

Database systems

Another key to knowledge-based systems, and most other computing applications, is the database, the massive store of knowledge a computer needs to do its job. In the early days of computers, data was stored on the familiar reel of magnetic tape, but data retrieval was slow.

Even modern bulk storage methods using high speed disks are slow relative to the speed at which a computer computes. Databases are also growing to the extent that entering all the data may take months. Accessing the data can take hours because of the volume. MCC's database program aims to get more information in and out faster. At MCC, database machines will be built, said program director Eugene Lowenthal, formerly with Intel Corp., and each machine will rely on both parallel processing and VLSI circuits.

Although Lowenthal had hoped a single database structure could satisfy the needs of all MCC computing programs, it now appears, he said, that two different types of databases will have to be developed. The first—accounting for 75 percent of the program's effort—is a "logic database," akin to conventional ones. In these, data will be entered and retrieved through logic (Prolog-like) assertions. The second is called an "object-oriented database." It will make use of Lisp (symbolic) processing, to give the programmer maximum control over computation.

Lowenthal said MCC would not develop distributed database systems because many outside groups are working to perfect them. However, although logic programming optimized for small databases that fit into a computer's central memory is being developed in industry, "no one has ever built a compiler for large databases that reside on disk," Lowenthal noted, adding that this will be done in-house.

The logic database is to implement a new language that MCC

experts developed, called LDL, for logic data language. It is an upgraded version of the language Prolog, Lowenthal said.

In several years, Lowenthal hopes to combine the two databases. Ideally MCC's database program should be completed before the AI-KBS program even begins, Lowenthal said, because databases are "lower on the food chain." As a result, he said, some rework will have to be done when the knowledge-based prototypes are built.

Lowenthal noted that Japanese researchers are not developing logic compilers or object-oriented databases, essential links between databases and knowledge-based systems. He said this was a flaw in Japan's approach to artificial intelligence work.

Lowenthal indicated that for MCC shareholders the results from his program could yield huge income. The LDL language "could become a general purpose programming language," he said, making conventional databases much more powerful and therefore more marketable. And MCC will build several database machines orders of magnitude more efficient than today's, which could be adapted by shareholders for sale in mass markets.

Human factors

In all three of the computer programs and in the software and CAD programs, emphasis has been placed on improving the efficiency with which the programmer, designer, data entrant, and end user interact with the computer. For this, all of the programs rely on the human factors group.

A wide range of techniques are being explored, including voice recognition, natural language processing, and three-dimensional graphics. But the program director, Raymond Allard, formerly with Control Data Corp., is starting with the basics. During *Spectrum's* visit, his researchers were observing through one-way glass how people interact with computer terminals. Psychologists on Allard's team are developing cognitive models of the interaction; later, behavioral studies will guide interface design and help evaluate new software.

Linguists are helping to develop more powerful natural language techniques, concentrating on multisentence dialogues. They and the behavioralists are working with AI researchers and systems designers to create a user's assistant—some combination of technologies that prompts users, accepts various inputs, and answers questions. It will advise and coach users and adapt to each one. Computer design engineers are to incorporate speech recognition and image handling devices.

Allard is also working to move beyond icons, the graphic symbols on a terminal that represent functions. Although perhaps they are valuable to unaccustomed computer users, they slow a familiar user's progress and are becoming too numerous to be easily distinguished from one another. Rather than icons, objects may be shown in realistic settings and in abstract diagrams. Allard said he might use three-dimensional graphics so a user could ''fly'' through an instruction sequence, as if the user were in an airplane passing through gaps in clouds that delineated various sequence paths.

By the end of 1988, Allard hopes to demonstrate an intelligent user-interface management system with general software building blocks and tools to customize them. Once this system has been designed, advanced interfaces will be built.

Allard also suggested that prototypes from his program might give shareholders in his program major advantages in the commercial market. ''Ultimately,'' he said, ''computer products sell based on which ones potential users like the most and which are easiest to use. That means human factors.''

Reprinted from *IEEE Spectrum*, Mar. 1986, pp. 76–82.

17/THE STATE OF SCIENCE AND TECHNOLOGY IN GREAT BRITAIN

David Phillips

Science tends to grow exponentially. Money to pay for research does not. Although the rate of growth in the civil science budget in Great Britain began to tail off in the 1960s (Fig. 17-1), it is only in recent years that the real strains on the system have begun to show. The strains are reflected in the contrary positions taken on science funding: Government officials argue that their stated policy of maintaining level funding for science has been upheld, whereas the group of researchers mobilized in 1985 under the banner "Save British Science" believes a once great scientific tradition is facing calamitous decline.

From the perspective of the Advisory Board for the Research Councils, which each year frames proposals for the government science budget covering five research councils, the two views are not inconsistent. The research councils are the main granting agencies for academic and academically related research in Great Britain. The secretary of state for education and science almost always accepts the advisory board's advice on the allocation of funds to the councils, even if the board's overall budget demands are not met. Thus, in 1986–87, the total £614.6 million ($891.1 million) science budget of the Department of Education and Science was divided as follows:

- Science and Engineering Research Council—£315.5 million ($457.5 million)
- Medical Research Council—£128.3 million ($186 million)
- Agricultural and Food Research Council—£52.7 million ($75.4 million)
- Natural Environment Research Council—£70.3 million ($101.9 million)
- Economic and Social Research Council—£23.6 million ($34.2 million)

In addition, there were smaller allocations to the Royal Society and

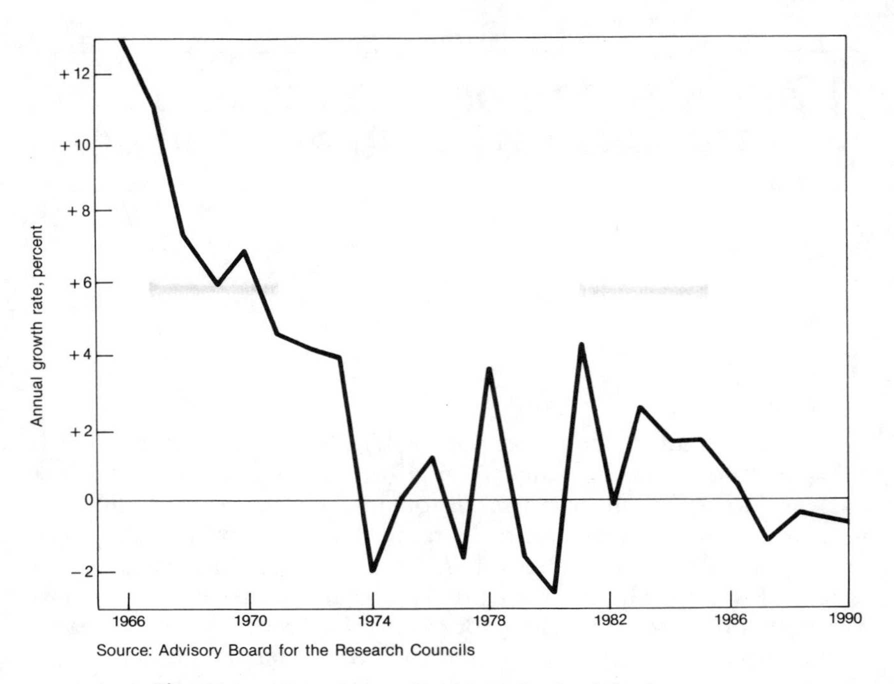

Fig. 17-1: *Growth rate, British civil science budget.*

other entities. The five research councils can point to expansions in scientific opportunity. The Science and Engineering Research Council, for example, which funds particle physics and astronomy as well as basic chemistry, biology and mathematics, and engineering research, is also responsible for such increasingly important areas as computer science, molecular electronics, and new materials. Finding the funds to promote such research, however, is increasingly difficult.

The cash value of the science budget has been roughly maintained in real terms over the last five years, although often by last minute additions grudgingly granted. For several reasons, however, cash parity is not sufficient to maintain the volume of research. Some of these are peculiar to Great Britain, some exist worldwide. Together, they add up to a research system in rapid change, trying to promote important new areas of science while balancing the books.

Some of the roots of British researchers' disquiet lie in key differences between funding arrangements on the two sides of the Atlantic. Great Britain operates a so-called dual support system in which most government-backed civil research is supported cooperatively by the 40 or so British universities and the research councils. (The major source of university funding is an annual government

TABLE 17-1
BRITISH GOVERNMENT R&D EXPENDITURES (IN £ MILLION, CONSTANT 1983–84 PRICES)

	1981–82	1983–84	1985–86	1987–88 planned
Civil departments	953.6 ($1,382)	907.3 ($1,315)	930.5 ($1,348)	787.4 ($1,141)
Research councils	469.3 ($680)	479.8 ($696)	494.6 ($718)	491.8 ($713)
Universities	539.0 ($781)	551.0 ($799)	530.6 ($770)	520.0 ($754)
Total civil	1961.9 ($2,845)	1938.2 ($2,810)	1955.8 ($2,836)	1799.2 ($2,609)
Total defense	1942.8 ($2,817)	1984.0 ($2,877)	2181.3 ($3,162)	2181.3 ($3,162)
Total	3904.7 ($5,662)	3923.0 ($5,688)	4137.0 ($5,999)	3980.6 ($5,772)

Source: *Annual Review of Government Funded R&D,* Her Majesty's Stationery Office, 1985.

allocation of £1.2 billion ($1.7 billion) for higher education.) Table 17-1 shows that universities and research councils together account for just over a quarter of total government funded R&D. Table 17-2, taken from a study performed for the Advisory Board for the Research Councils by the science policy research unit at Sussex University, shows government spending in different fields in 1982. In 1982, in absolute terms, Great Britain was behind all the other countries studied except the Netherlands. The relative position of Great Britain has almost certainly worsened since then [1].

The University Grants Committee distributes the annual £1.2 billion government allocation for higher education. In theory, part of the block grants to individual universities from the committee goes toward basic scientific equipment and laboratory overhead and pays for some research. Larger projects are then funded by grants from the research councils or elsewhere. In this system, which is both more centralized and more dependent on government funds than that in the United States, the nation's overall research capability depends crucially on general university funding.

Furthermore, while research council budgets have stayed roughly level since 1981, university funds have been cut. The estimated research contribution from the overall university grant declined by some 3.5 percent in real terms between 1981–82 and 1987–88, according to official figures. However, most believe the decline has been greater because universities have protected teaching at the expense of research as budget pressures have grown.

On the research councils' side, the purchasing power of the money

TABLE 17-2
COMPARATIVE FUNDING FOR ACADEMIC RESEARCH, BY FIELD, 1982
($ MILLION)

Field	Great Britain	West Germany	France	Nether-lands	United States	Japan
Engineering	304.6 (15.8%)	409.6 (12.4%)	205.6 (8.0%)	90.0 (10.1%)	1,135.8 (12.1%)	683.4 (21.6%)
Physical sciences	415.7 (21.6%)	835.5 (25.3%)	843.8 (32.6%)	175.5 (19.6%)	1,416.2 (15.1%)	451.5 (14.3%)
Environmental sciences	127.8 (6.6%)	153.8 (4.7%)	119.2 (4.6%)	21.8 (2.4%)	591.9 (6.3%)	101.2 (3.2%)
Mathematics and computing	86.0 (4.5%)	114.6 (3.5%)	136.4 (5.3%)	32.6 (3.6%)	281.3 (3.0%)	72.5 (2.3%)
Life sciences	662.4 (34.4%)	1,192.4 (36.2%)	935.2 (36.2%)	345.8 (38.6%)	4,629.8 (49.4%)	1,046.8 (33.1%)
Social sciences (includes psychology)	112.3 (5.8%)	163.9 (5.0%)	119.2 (4.6%)	98.2 (11.0%)	592.2 (6.3%)	132.0 (4.2%)
Professional and vocational	96.7 (5.0%)	175.4 (5.3%)	60.0 (2.3%)	57.3 (6.4%)	269.3 (2.9%)	334.2 (10.6%)
Arts and humanities	119.7 (6.2%)	228.4 (6.9%)	165.6 (6.4%)	72.6 (8.1%)	286.7 (3.1%)	344.0 (10.9%)
Multidisciplinary	—	23.2 (0.7%)	0.5 (0.0%)	0.8 (0.1%)	169.4 (1.8%)	—
Total	1,925.4 (100.0%)	3,296.6 (100.0%)	2,585.4 (100.0%)	894.9 (100.0%)	9,372.7 (100.0%)	3,165.6 (100.0%)

Source: Science policy research unit, Sussex University

granted has also declined because of increases in salaries and retirement benefit costs. The international weakness of the pound has made it harder to pay subscriptions to overseas scientific organizations, such as CERN (European Center for Nuclear Research) in Geneva, and there have been perennial increases in the cost of laboratory instrumentation—dubbed the ''sophistication factor'' in Great Britain. Overall, the Advisory Board for the Research Councils estimates that the volume of research funded through the science budget, as opposed to general university funding, will shrink by more than 10 percent over the next decade [2].

This decrease is in part a sign of an inevitable transition from the exponential growth of government funded research, which has characterized the economies of members of the Organization for Economic Cooperation and Development since World War II, to spending levels linked more tightly to movements in gross domestic product—the transition that science historian Derek DeSolla Price foresaw more than 20 years ago. But if so, Great Britain's relatively poor economic performance has brought about this transition ahead of other countries. At least until 1986, our main economic competitors were still increasing their research budgets. In consequence Great Britain lags on most international indices of research spending, especially for civil research. There looms a painful double bind. Our economic weakness makes it hard to compete internationally in science. But a failure to compete may perpetuate that weakness.

As chairman of the Advisory Board for the Research Councils, I believe that the board must argue for real budget increases. At the same time, though, the board must convince researchers that they can expect something much closer to steady state funding than they became used to in the 1960s, when the budget was still doubling every 10 or 15 years. In addition, any extra money the board can win from government will be in return for stronger assurances that the money will be spent wisely. Advisors such as those on my board have to find new ways of answering the key policy questions: How do we back the best quality research? How do we promote selectivity between program areas and institutions? How do we strike a balance between funding people and instrumentation? How do we ensure that strategic research fields are selected with an eye to future economic potential? In short, how do we select and structure our research portfolio to justify our claims on the taxpayer?

Selectivity in research support is not new in Great Britain, as befits a country that accounts for only around 5 percent of global R&D. As long ago as 1970 the Science and Engineering Research Council—the largest spender—announced a policy of selectivity and concentration in research grants, in response to a decline in the annual rate of growth of its budget. Even then, in many areas, 10 percent of grant applicants received half the available money, although the selectivity then applied was more in terms of institutions than disciplines.

However, the demand for selectivity is undoubtedly increasing in response to the new pressures, notably on the university side of the system. The University Grants Committee is moving toward designating sheep and goats among its flock of 42 institutions, at least as far as research goes, with the consequent redistribution of funds. According to Peter Swinnerton-Dyer, chairman of the committee, "If the best research is to be adequately supported, the less good research will suffer more: but this is better than a policy of inadequate support for everybody [3]."

The committee now allocates part of each institution's grant specifically on research-based criteria, although universities are still free to spend the money as they wish. The teaching component, which accounts for most of the committee's £1.2 billion budget, is fixed solely by student numbers, as before. The research component has four subsidiary elements:

- A sum equivalent to 40 percent of an institution's income from research council grants, to make dual support more explicit
- A sum related to student numbers and academic staff, for general research overhead
- A sum related to external contract research income
- A sum determined by judgment of an institution's research strengths—as weighed by individual subject committees with advice from research councils and others.

One result has been, in effect, a ranking of British universities by scientific department and discipline—although this was not the University Grants Committee's intent. There has been much unhappiness in many departments that failed to achieve a "star" in the committee's research rankings, and the recently published results will likely affect student choices to some extent. In May 1986 Keith Joseph, then secretary of state for education and science, told the House of Commons that the overall funding allocation exercise was a "landmark" in university policy. The committee's allocations for each discipline were fixed before the calculations about institutional distribution were made, and the total sums moved around will be relatively small in the first years. Nevertheless, the long-term effect will be to shift the cash to universities that are successful at winning external grants and that the committee judges to be strong research centers—presumably the same set. The procedure marks a step toward a possible stratification of universities into high level research centers and institutions mainly concerned with teaching, much more like the existing U.S. system.

Meanwhile, the research councils are grappling with their own agenda for selectivity. For some of them this has involved large-scale restructuring of programs and heavy staff losses. The Agricultural and Food Research Council, for example, has shed 2,000 staff from the original total of 6,000 employed in its own institutes while managing a shift from farming to food processing research on a declining overall budget [4].

For others, such as the Science and Engineering Research Council, a larger proportion of university grant applications has been rejected, and spending on central facilities for university research has been cut.

The research councils and the advisory board are also interested in selecting areas of strategic research for support in those scientific disciplines most likely to underpin future industrial expansion. The

Advisory Board for the Research Council's sister committee, the Advisory Council for Applied Research and Development, which advises the prime minister on R&D and innovation policy but has no specific budgetary or executive responsibilities, published a report earlier in 1986 calling for an effort to identify "exploitable areas of science." The advisory council argued that strategic research, concerned with both advancing fundamental knowledge and developing new technological possibilities, is assuming more importance in a climate of stronger international industrial competition.

The council's report suggested that the United States is probably the only country that can afford to fund basic research across the board, treating the commercial exploitation of science as a random process. For the less healthy, judicious selection is essential. The council looked to Japan's extensive system for scientific and technical forecasting for inspiration in outlining a process to help Great Britain formulate national R&D priorities geared to industrial growth. The intent would be to try to identify exploitable areas of science, defined as areas "in which the body of scientific understanding supports a generic (or enabling) area of technological knowledge; a body of knowledge from which many specific products and processes may emerge in the future." In other words, the advisory council would go beyond the broad emphasis—now shared by many countries—on biotechnology, information technology, and new materials and would select promising areas in more detail. However, the council also recognized that Great Britain does not have the systematic data collection needed for such an effort. If the preliminary studies begun in the wake of this report bear fruit, which will depend partly on industry putting up the bulk of the money, they could have important results for development of British science policy [5].

To round off this brief description of moves by research funding agencies toward greater selectivity, I must mention the further pressures—also familiar in the United States—that force those who hold the purse strings to think deeply about the distribution of funds. A striking example of these pressures comes from the United States, in fact—namely the discussion of instrument needs in the National Academy of Sciences survey entitled *Opportunities in Chemistry* [6].

There are important implications for Great Britain in the academy panel's conclusion that a university chemistry department with the broadest research capability must maintain six out of a possible eight state-of-the-art laboratory instruments—and that at current U.S. funding levels only 20 out of 198 Ph.D.-granting institutions in the United States can expect to operate at this level [7].

The academy report also envisages a further 40 second-tier and 60 third-tier institutions whose financial needs are part of a general case for additional funding. This analysis points up issues that British research administrators have scarcely recognized. It is widely assumed in Great Britain that every university worthy of the name must have a chemistry teaching department and that every teaching department must do some front-rank research. The Advisory Board for the Research Councils must now ask if our relatively small funding can support anything like the same number of research departments in chemistry as there are universities. If not, there will be hard decisions on location and selection of the real research centers. These questions, and the analogous set in other disciplines, raise long-term, strategic issues that the advisory board will address later this year and in 1987. Great Britain, which has long muddled through without any explicit overall science policy, may now need something like a national plan for science—at least in the academic sector.

A plan for science, if such a thing is possible, will be merely a bureaucratic exercise unless its authors address how to maximize the benefits from research. It is often alleged that Great Britain is brilliant at basic research but inept at translating ideas into hard cash—a perception manifested in attitudes like those of U.S. companies who keep abreast of basic science by "renting a Brit." At home, many compare our record unfavorably with our competitors in Asia; in caricature, the British invented radar and the Japanese adapted the same technology to sell microwave ovens to the world.

In reality, of course, the British preoccupation with improving links between academic research and industry is shared today by many other countries, including Japan. And we have some very good examples of how to succeed, as well as some less auspicious cases. The growth of high technology industry around Cambridge University does not exactly rival that of Silicon Valley, but the area can boast more than 300 start-up companies over the last 25 years—the vast majority in the last decade. The lessons we have learned from both the successes and the failures are now beginning to feed through into policymaking [8].

My own feeling is that much earlier policy thinking, in Great Britain at any rate, was hampered by allegiance to a misleadingly simple linear, or assembly line, model of the innovation process. This model, in which inputs into basic research lead through a sequence of stages—applied research, development, design, and marketing—to product or process innovations, was sold to the British government after World War II, just as Vannevar Bush sold it to the Truman administration. But it does not bear close examination.

Great Britain has expended much effort over the years on trying to make its exploitation of science fit the linear model, with often disastrous results. Organizations such as the former National Research Development Corporation had a disappointing history of unsuccessful attempts to introduce innovations based on clever scientific ideas. For example, it turned out that there was no significant world market for the Hovercraft.

We are now nearer to recognizing a more complex, interactive model of innovation, in which contact between academic research and industry is needed at all levels to create the most fruitful coupling between the research base and the market. One implication is that there is no single most favorable institutional model for expediting profitable innovation. And Great Britain is currently home to as diverse a range of experiments in technology transfer as any country. We have university science parks, industrial liaison teams, private sector technology transfer agencies funding the development of promising ideas, computer data bases of academic research, and collaborative research schemes linking research councils, universities, government departments, and industry. Much of this is new and untried. As a working group of the Advisory Board for the Research Councils reported in May 1986, "A ferment of activity is in progress, the results of which will not be known for some years [9]." The rationale for most such schemes is to bring academe and industry closer together and give each side more knowledge of the other's priorities, rather than simply to speed up the movement of ideas from the laboratory to the factory.

This is all promising, but there are still tensions among different priorities. Government pronouncements sometimes give the impression that universities should do more research of direct, immediate relevance to industry—and the pressures on college funds have certainly induced some campus researchers to move this way. But there are real fears that this shift could undermine the long-term future of basic research.

Government is more likely to heed this message from industry than from bodies such as the Advisory Board for the Research Councils (although the board has industrial members). And there are encouraging signs of British industrialists rallying support of public funding for basic research. For example, John Harvey-Jones, chairman of ICI, the British chemicals giant, put the point uncompromisingly in a broadcast lecture earlier in 1986: "We are in danger of failing to recognize that our science is a major strength, is a capacity uniquely suited to the modern world and ought to be a cornerstone of future competitiveness. The Government must ensure that our national intellectual and scientific assets are effectively sustained by its support of fundamental research [10]." Statements like this are commonplace in the United States, but it has hitherto been relatively rare for British business to recognize the importance of novel science and technology.

The domestic picture in Great Britain is thus one of rapid change. Nevertheless, in a small country, policy for science tends to be more and more shaped by developments abroad, and this is especially true for Great Britain, which often is torn between throwing in its lot with European partners and forming stronger links with the United States. Current issues such as the future of high-energy particle physics, participation in the Strategic Defense Initiative (SDI) research program, and establishment of the pan-European Eureka (European Research Coordination Agency) civil research program exemplify this conflict.

In some areas Great Britain's focus is firmly European. In particular, officials of the governments concerned are moving to internationalize Europe's admirable facilities for condensed-matter research. The Science and Engineering Research Council's new neutron source, called ISIS, is to be offered to French, Italian and, probably, West German researchers, in return for British researchers having access to the planned European synchrotron radiation facility in France. ISIS, at the Rutherford Laboratory in Oxfordshire, is the world's most intense source of pulsed neutrons and will be widely used by biologists, chemists, and physicists investigating everything from surface effects to protein structures. It will complement the X-ray beams from synchrotron sources. Great Britain is also a founding member of the European Space Agency and takes part in such European Economic Community programs as ESPRIT (European Strategic Program for R&D in Information Technologies), as well as in numerous smaller scientific exchange schemes. In particle physics Great Britain has a particular interest in future relations between the cooperative European laboratory at CERN in Geneva and similar research centers in the United States because the British government is seeking a 25 percent reduction in its contribution to CERN. On the face of it, particle physics is ripe for the final stage of evolution toward intercontinental, if not global, collaboration to construct the next generation of accelerators. However, there is a widespread perception in Europe that the spirit behind current U.S. proposals for the Superconducting Supercollider is emphatically competitive, rather than cooperative—a spirit sharpened by the spectacular successes at CERN in the early 1980s, when the W and Z particles were characterized.

In my view all governments should now ask why they support high energy physics. If it is for pure scientific interest, wider cooperation must be the way to go. The promise of applications in fields such as energy technology should be no bar to collaboration—any more than in further development of fusion power. If, however, there is now military interest in particle beams and new accelerator technology, wider collaboration will be harder to achieve. The spirit of CERN is certainly opposed to any weapons-related applications of the laboratory's expertise.

That opposition is also reflected in attitudes among British scientists to participation in SDI research. Although the British government was the first to agree formally to collaborate with the United States in this area, a significant number of British researchers in key disciplines—notably computer science and physics—have opposed any involvement in what they regard as a dangerous and misguided program. This, of course, will not deter others from applying for funds. When they do, there will still be residual British concerns about whether Great Britain can retain control of the intellectual property rights accruing from SDI research—and also about the implications for the balance of the British effort in a few crucial fields. The needs of the SDI program in software engineering, for example, may not be the same as those of civil industry.

In addition, many feel it would be unfortunate if the prospect of support for British researchers from the SDI budget drew an even larger proportion of Great Britain's best R&D people into defense-related work. Great Britain is the only Western country that spends anything approaching the U.S. percentage of its government R&D budget on defense work—more than 50 percent at present. And the fact that most of the basic research associated with British defense programs goes on in Ministry of Defense establishments, in contrast to the U.S. model of university administered laboratories, has limited private sector commercial exploitation of this work.

A number of related efforts are under way to break down the barriers around British defense research. They include a company specially formed to assist commercialization of selected promising ideas from Ministry of Defense laboratories, a joint research councils/Ministry of Defense grants program for university research launched in 1985, and discussions about joint planning of large experimental facilities involving civil and defense research programs. But none of these efforts changes the fact that while Great Britain's civil science is losing funds—and international standing—the country continues to spend more than £2 billion ($2.9 billion) a year on defense R&D. True, much of this is development work that might be classified as weapons procurement, but most members of the civil science community believe the defense program must be trimmed to give a more balanced set of national R&D priorities.

As suggested already, those priorities will also have to be framed while keeping a close eye on European R&D. In research programs designed to produce innovations for industrial development in the relatively short term, for example, the full fruits can be realized only by collaboration between firms that can address the European market as a whole—thus benefiting from the same advantages as U.S. firms that can launch their initial sales efforts in a large, homogeneous domestic market.

Thus, there is a growing British consensus that a second phase of the joint academic/industry program for artificial intelligence and computer science research must be "Europeanized" when the first five years and £200 million ($290 million) of state funds come to an end in 1988. This program, launched in answer to the Japanese fifth-generation computer effort, is known as the Alvey program—after John Alvey, the author of a report to the government that proposed it. This program is designed to promote research and industrial development in advanced information technology and computer science, and it brings together universities, research councils, companies, and the departments of education and science, trade and industry, and defense. But it is restricted to British companies, unlike the similar European Economic Community program ESPRIT, in which firms from two or more countries are involved in each project. A further phase of the Alvey effort will have to be linked closely with ESPRIT.

Related to this aim is the broader, and less well-defined, European initiative known as Eureka. This is shaping up to be strongly market driven, partly at Great Britain's behest, and it will be a vehicle for European firms to develop collaborative R&D with the blessing and, possibly, some financial encouragement from participating governments. Eureka is supposed to promote the application of new technologies in European-wide markets and overcome the disadvantages of markets divided among relatively small countries. The official British position at present is that there will not be new money for Eureka, but British firms will be able to put Eureka projects forward for support under existing schemes to promote industrial innovation supervised by the British Department of Trade and Industry.

SDI, seen not as a defense program but as a massive U.S. government subsidy for new technologies of significant industrial potential, again looms in the background of Eureka. In this sense Eureka was conceived in part as a response to SDI, although the Eureka proposals have also crystallized a feeling that Europe must get its own act together in high technology R&D.

In any event the core technologies that figure in both programs overlap heavily, especially in such areas as computer architecture, software, sensors, and new materials. Great Britain's position of signing up for both now looks more likely to be shared by other major European countries following the West German agreement on SDI and France's recent change of government.

A final, highly visible issue in Great Britain, and one that affects British thinking on science matters, is the outflow of talented scientists to other countries—most often the sunnier parts of the United States. The so-called brain drain has been an important part of the case for more civil research funds in the last couple of years, as the flow appears to be increasing. Great Britain already contributes more scientists and engineers—around 1,000 a year—to the U.S. technical work force than all other European countries put together, and there is serious concern in Great Britain about the effects of this in key fields. The Advisory Board for the Research Councils is studying the volume and character of the outflow to see what may be done to induce the more able researchers to stay home.

British institutions cannot possibly compete with the salaries offered by some U.S. employers, but the advisory board may be able to cater to those prospective emigrants whose concerns focus more on research opportunities and standards of laboratory instrumentation. For example, a special addition of some millions of pounds to the universities' equipment budget in 1985 was allocated to a very few groups, to ensure they had state-of-the-art instruments. We may see more of this selective extravagance as a matter of advisory board policy.

Anyone contemplating the state of science and technology in Great Britain must consider where the next generations of researchers will come from, especially if we continue to lose so many early in their careers. Science can reproduce itself only if there is a steady supply of aspiring researchers coming out of the schools and seeking an apprenticeship at the laboratory bench. This is a serious cause for concern in Great Britain. Demands from industry for highly qualified scientists and technologists are revealing shortages in such key skills as computing, which raise salary levels beyond universities' means.

Efforts to meet these shortages are being hampered by the condition of British primary and secondary education. School teaching does not have high status in Great Britain, and a loss of morale among teachers combined with low salaries has drastically curtailed recruitment of the mathematics and physics teachers essential to bringing students up to university entrance level in science. Government policy calls for curriculum changes that will ensure that all students take some science subjects, but they may well be supervised by teachers unqualified in science unless measures such as use of differential salaries can raise the numbers of qualified teachers. Great Britain appears on the verge of a renewed debate about educational standands, with special emphasis on the quality of technical and vocational education, but the results will take a long time to affect the numbers of qualified teachers.

Related to this is a wider debate familiar in the U.S. context about the level of public understanding of science, or scientific literacy. This concerns many scientists partly because of worries over economic performance and attitudes regarding new technology, but also because of some currents of public opinion that come closer to the laboratory. Examples of the latter include strong opposition to *in vitro* fertilization research—a bill that would have outlawed all experiments on human embryos went through several stages of debate in Parliament before being filibustered and so defeated—and criticism of experiments using animals. Militant animal liberation groups have raided laboratories, threatened researchers, and damaged the researchers' property up and down the country. They regard new British legislation regulating animal research as a charter for vivisectionists.

British researchers naturally regard these currents of opposition as stemming from a failure to appreciate the merits of their work. A 1985 report from the Royal Society rightly concluded that education must be the main target for action on this front [11]. But the Royal Society also urged working scientists to take more responsibility for explaining their research to lay audiences. This is a difficult responsibility to place on researchers already trying to keep pace with fast developing fields on tight budgets. However, perhaps if it comes to be taken more seriously in Great Britain, as it already is in the United States, making the case for science funding would also become easier. That in turn may help achieve long-term security of funding instead of an annual uncertainty about next year's budget.

References

[1] B. R. Martin and J. Irvine with N. Minchin, *An International Comparison of Government Funding of Academic and Academically Related Research*, ABRC Science Policy Studies No. 2. Brighton, England: Sussex University, Science Policy Research Unit, Oct. 1986.

[2] Department of Education and Science, *Science and Public Expenditure 1986: A Report to the Secretary of State for Education and Science from the Advisory Board for the Research Councils.* London: ABRC, July 1986.

[3] P. Swinnerton-Dyer, "Weighing Out the Pots of Gold," *Times Higher Educational Supplement*, Nov. 15, 1985.

[4] Agricultural and Food Research Council Corporate Plan, 1986–1991. London: AFRC, March 1986.

[5] *Exploitable Areas of Science: A Report by the Advisory Council for Applied Research and Development.* London: Her Majesty's Stationery Office, 1986.

[6] National Research Council, *Opportunities in Chemistry.* Washington, DC: National Academy Press, 1985.
[7] National Research Council, *Opportunities in Chemistry,* pp. 303–5.
[8] *The Cambridge Phenomenon: The Growth of High Technology Industry in a University Town.* Cambridge: Segal Quince and Associates, 1985.
[9] Advisory Board for the Research Councils/Department of Education and Science, *Report of the Working Party on the Private Sector Funding of Scientific Research.* London: ABRC, May 1986.
[10] J. Harvey-Jones, "Does Industry Matter?" *The Listener,* Apr. 10, 1986.
[11] *The Public Understanding of Science: Report of a Royal Society Working Party.* London: Royal Society, 1985.

18/THE STATE OF SCIENCE AND TECHNOLOGY IN GREAT BRITAIN: AN ACADEMIC'S VIEW

Hans Kornberg

The ancient statutes of Christ's College—one of the thirty or so autonomous entities that make up the University of Cambridge—charge each Fellow to promote the college as a place "of religion, education, learning and research." Although the first of these categories may not be unambiguously relevant to the state of science and technology in Great Britain, the remaining three undoubtedly are and neatly encapsulate the essential qualities of a university.

More than any other educational institution, a university should be dedicated to the advancement of learning through research. Important also, but secondary, it should educate the scholars of the future and, in particular, should train them in the philosophy and methods of inquiry. Third, it should, through various means of communication, pass on the fruits of that research both to the wider community and to those who wish to apply the knowledge gained to the common good, which includes the generation of public and private wealth. It follows that I believe a teacher of science in a university must be actively engaged in research. If he is not, he cannot adequately teach his students how nature can be persuaded to yield up her secrets, or how they can critically evaluate and interpret the work of others.

I make these perhaps overly idealized and certainly overly simplistic remarks because they provide my frame of reference, against which I can measure my reaction both to the current scientific scene in Great Britain and to the direction in which it is moving.

From his official vantage point, David Phillips has outlined the various ways in which the British government is squeezing both legs of the dual support system and is thus obliging universities simultaneously to contract the research they do and to change its intensity and orientation. Many of these constraints have barely had time to make themselves felt, particularly in departments such as mine (biochemistry) that have enjoyed fairly generous support from the research councils in the past. However, even here the signals are unmistakable and ominous.

For example, the University Grants Committee has just announced that its grant to the University of Cambridge for 1987 will be only 0.7 percent larger than that for this year. Since inflation currently runs at more than 3 percent, some drastic belt tightening will be urgently required. Thus, one of two academic staff posts currently vacant in my department cannot be filled; the other one can, but only at the lowest possible salary level. And, since no diminution in student numbers is likely to occur in the near future, the remaining members of staff will have to devote less time to research and more to undergraduate instruction. There will also be very little money to replace obsolete and outworn equipment.

The other leg of the dual support system is no stronger. A close colleague and I applied to the Science and Engineering Research Council for two research grants. Both were graded "alpha," the top category in the peer review system. But one was funded only to an extent that inevitably caused the money to run out long before the end of the project, and the other was not funded at all. This, in turn, had several consequences, some of which may even be beneficial in the long run. When we failed to attract research funds from the usual sources, we undertook intensive efforts to open up new ones, particularly from industry. The response has been encouraging in that several major British companies have been willing to support fundamental work that is not likely to be of direct benefit to them. Such support, however is on a much smaller scale than that needed to sustain a research group of significant size, even in my relatively inexpensive discipline. None of the six postdoctoral students who worked with me two years ago now holds a position in any British university.

The inability of the research councils to fund more than a fraction of even the grants rated alpha is not due only to the absolute shortage of money. It is also due to the diversion of resources from areas of fundamental inquiry into areas deemed to be of more immediate industrial relevance, and to the concentration of support on programs that the research councils themselves designate as priority areas.

It is this latter attempt to grade areas of research that fills me with the greatest apprehension. Even if the quality of those who do the grading were beyond question, how many of the most important discoveries of the past decade have unpredictably come from then unfashionable disciplines (and equally unpredictable universities)? The University Grants Committee also has begun to give preferential funding to those universities that have best succeeded in obtaining research council support. We see, therefore, that universities will be placed either on an escalator that will carry them onward and reasonably upward or on one that will bring an inevitable deterioration in the scale of their facilities and resources.

This potentially disastrous process, which will ensure that bright young people who do not happen to find themselves in universities on the up escalator will find it very difficult to undertake research in science, is based on a misapplication of a concept advanced in 1970 by the Science and Engineering Research Council, of which I was then a member. Obviously, a country the size of Great Britain cannot sustain more than a few "big science" projects—such as those in high energy physics, radio astronomy, or space exploration. The council therefore provided central facilities that were intended to help service both large international laboratories, such as CERN (the European Center for Nuclear Research) in Geneva, and the smaller university groups that needed to carry out work in them. The research council also attempted to set up central facilities for those "little sciences" that were growing in complexity of equipment (and, consequently, expense), but they proved not to be wholly successful.

It was still easier to obtain funds to buy one's own machine than to travel to some distant center in the hope of obtaining sufficient time on the machine provided there. In my opinion, however, there is now an overwhelming case for reviving and reviewing this idea. If sophisticated items of equipment were placed in special well-serviced centers attached to appropriate university departments, with the proviso that all qualified researchers should have access to them, the down escalator may yet be dismantled.

Colleagues abroad have noted that there are many scientists in Great Britain (and I am among them) who believe that current official science policies are dangerously mistaken and who are attempting to change them. But even occasional bright interludes, such as a recent sizable injection of government funds into the basic research budget, do not provide much cause for optimism. I am reminded of a monologuist on the radio who, in the darkest days of World War II, ended his recital of unmitigated disasters thus:

"And then I heard a voice say: 'Cheer up! Things could be worse!' So I cheered up. And they were...!"

19/THE STATE OF SCIENCE AND TECHNOLOGY IN GREAT BRITAIN: AN INDUSTRIALIST'S VIEW

Derek H. Roberts

For Great Britain to afford the improved way of life that its citizens demand, it must strive to provide sufficiently good value in its manufactured goods and services that the world will be responsive to British engineers and salesmen—and women—when they set out to conquer global markets. It is science-based technology that will enable this to happen, but of course it is the quality of British scientists, engineers, technologists, and managers that will make it happen.

There are many ways in which the universities should be and are of considerable relevance to industry. The most important of these is the way in which they educate and motivate the young men and women who will shape the industry of tomorrow. For this educational and motivational role to be successful, the quality and motivation of the academic teaching staff is paramount. In my view the most important justification for maintaining a strong basic and strategic research base in pure and applied science and engineering is to provide the environment in which undergraduate and postgraduate education can flourish.

Research provides an important bridge linking industry and academia. Collaborative research, across this bridge, is an essential aspect of such early initiatives by the Science and Engineering Research Council as Cooperative Awards in Science and Engineering (CASE) studentships and the cooperative research grant scheme. More recently, the Alvey and ESPRIT (European Strategic Program for R&D in Information Technologies) programs, to which David Phillips refers, all place—quite rightly in my view—considerable emphasis on industry-academic research collaboration.

Industrial research is concerned with the creation of opportunities associated with new products, new manufacturing processes, and new markets. In the absence of an adequate research foundation, subsequent product development and launch will be more risky, more expensive, and take longer—which implies a higher degree of commercial uncertainty.

Industrial success increasingly depends on technical and market innovation. To this end the total British R&D effort and its direction are vital to future commercial success. However, that part of the total R&D expenditure that is devoted to the maintenance of basic and strategic academic scientific research is quite small. If redeployed to expand applied research and development, it could at best have a marginal short-term benefit. On the other hand, such redeployment would damage the educational system and seriously undermine the mid- and long-term applied research capability of the country, since industry cannot be expected to maintain basic science.

In general, a company turns to the universities for support in areas that are complementary to that company's own capabilities. Well-managed, technology-oriented companies (whether large or small) will not therefore expect universities to carry out short-term product development for them.

In all of the above I use the word "science" to embrace pure and applied science and engineering. The "engineering versus science" argument is sterile in my view and totally ignores the relevance of science to industry. For example, it is not at all obvious to me that funds saved in "big science" such as high energy particle physics should be deployed in "engineering." Other areas of science may be better deserving in terms of the quality of the research, its benefit to the educational role of the universities, and even its industrial relevance.

Industry should play a more active role in lobbying the government to give greater recognition of the need for national priority investment in education—at all levels—and high quality scientific research. Similarly, industry should establish closer and better contacts with schools, colleges of all kinds, polytechnics, and universities and thus encourage our best young people to seek creative roles in wealth-producing industry.

Science and engineering disciplines, when well taught, should be recognized for their educational value. They should not be considered merely vocational training.

Academia and industry share a common goal—and have joint responsibility for achieving it—namely, improving the international competitiveness of the British manufacturing industry. Failure is unacceptable. It would lead to further increased unemployment and erosion of Great Britain's scientific and cultural base—and indeed of all that we admire in the British way of life. Success depends on partnership in research; on the encouragement and education of the nation's best young people as scientists, applied scientists, and engineers; and, increasingly, on the field of continuing eduation to equip people to cope with continuing technological change throughout their working lives. Academia's capability to contribute to this partnership depends on the health of basic science in our universities.

20/AN AMERICAN IN TOKYO: JUMPING TO THE JAPANESE BEAT

Daun Bhasavanich

The production line at one Westinghouse Electric Corp. factory has a sharp bend in it because of the uneven foundation of the building. When Japan's Mitsubishi Electric Corp., decades ago, took a license for the factory design, it modeled its entire operation after the Westinghouse factory—right down to the crook in the production line. Since 1923, when a licensing agreement between the two was struck, the companies have enjoyed a mirror-close relationship that survived even World War II. During the war Mitsubishi kept track of the royalties it owed Westinghouse and later paid up in stock shares. Westinghouse today owns about 2 percent of Mitsubishi.

In 1981, to transfer and acquire technology more effectively, Westinghouse began an engineer exchange program with Mitsubishi. For decades, the Japanese company had been sending a stream of engineers for one-year training internships at Westinghouse. Such exchanges are allowed under licensing agreements. Under the Westinghouse program, Daun Bhasavanich, an electrical and power engineer from Westinghouse's Research and Development Center in Pittsburgh, was one of those selected to work at Mitsubishi.

As preparation, he underwent a four-month training course in Japanese language, culture, and business communications. He completed his one-year stay at Mitsubishi in January 1984. In a highly revealing and personal account, he describes here his sometimes traumatic settling-in period and his major reorientation as he learned to work in the Japanese environment.

—Ed.—

Each day in Japan I arrived at work 15 minutes early to change into my uniform: plain blue trousers, a matching Eisenhower-type jacket embroidered with the company's three-diamond logo, a blue cap, and the company's black shoes. Work in the Circuit Breaker Engineering Division of the Mitsubishi

Electric Corp. (Melco) started promptly at 8 a.m., as it might at an electronics company in the United States. But there the similarities ended. Even before the workday began, stark differences between the two work environments were apparent.

For about 10 minutes before work started, I exercised to recorded music with my Japanese colleagues—young electrical and mechanical engineers. In my group we limbered up with two sets of exercises. First, there was five minutes of jumping rope outside our office building to popular Western tunes like the theme from the movie *Flashdance*. Then we moved inside for several minutes of calisthenics in front of our work desks to piano music, accompanied by instructions to stretch, inhale, or bend. Having warmed up, we began work at 8:00. There was no dawdling; to be even one minute late resulted in a tardiness demerit.

Every Monday morning our group held a section meeting. The section manager often gave business and technical information and then several employees were asked to speak. Popular subjects were recent product introductions from competitors of Mitsubishi. Two or three engineers or support technicians would talk for about five minutes each. The atmosphere was informal and informative. Everyone stood during the meeting, so it was naturally kept short.

During my early months at Melco, when I was far from fluent in Japanese, I reported on my first project—the design of an electronics-driven miniature device to operate inside a circuit breaker. The combination of the language barrier and the knowledge that the audience was standing kept my report succinct. The workday that followed was hectic to me at first. I shared an office with about 80 people. There was no professional secretarial help, but many women office workers performed manuscript copying and a variety of "household" chores, including serving tea.

Cleaning ceiling lights and drafting

I was given a desk and a drafting table, just like the others. I drafted, answered telephones, made measurements in the laboratories, wrote and filed notes—and cleaned the lights in the ceiling. Once I had to climb out of the office building onto a ledge, along with my engineer friends, to clean the outside of the windows. Another time we got down on our knees to weed lawn areas to prepare for a visit by an important Melco executive. While this may not seem an efficient way to use trained professionals, it certainly helps to quench any sense of elitism among company workers.

Our division manager, in charge of some 2,000 employees,

routinely spent two or three days a week on a two-hour walk through the offices and the factory. He wore the same uniform that his staff members did, although his rank afforded him a private office—the only one in the division. His walk helped convey a general sense that the local top management was approachable; workers like me had a chance for one-to-one discussions with him.

The lunch break was exactly 45 minutes long. Workmen on tricycles delivered company food, which cost about $1 a plate and was eaten at the work desk. The menu always included one box of rice and one side dish, like the fish of the day. No alternative menu existed, but no one brought brown bag lunches from home. Lunchtime games were limited to outdoor ping-pong, baseball, and miniature tennis. There was scarcely any time for a full game of chess.

Later, at 3 p.m., the familiar piano music for exercising was piped through the public address system, and just about everybody got up in front of his or her desk to stretch and flex. The exercise routine was completely standardized—one that Japanese had been performing since grade school; it would have been odd to see anyone start the first step with the wrong foot.

The workday ended officially at 4:45 p.m., but most people stayed on, some to 7 p.m., depending on the overtime quota. Everyone seemed to be working steadily at the desk all day, even in the overtime period. At first, when I would interrupt my work for a short break, my Japanese colleagues wondered if I felt ill.

Overtime, which pays an additional 30 percent above regular wages, is eagerly sought for several reasons. The extra pay helps meet the cost of living, which is even higher in Japan than in the United States. Moreover, the Japanese preference for gentle, personal persuasion, rather than bold management orders, slows decision-making on the job; overtime makes up the lost time to meet the schedule.

In some households where leisure comforts do not compare with those at the office, engineers prefer a late stay at the office to returning home early. A prestige factor may also be at stake—neighbors of the employee view overtime as a sign of how well the engineer is regarded by the employer.

Adapting to the work place

The office I occupied had no walls or partitions. Engineers shared technical files and storage space as well as company uniforms and company meals. On closer inspection, I found that the huge office was arranged like an orchestra pit, with the department manager, like the conductor of a symphony, at the head of the pit and the others fanned out radially from him according to seniority.

I arrived as an electrical and power specialist and was assigned promptly as an apprentice. Normally this period of apprenticeship—where one merely observes under the wings of each group member—lasts months. After I asked for an accelerated tour, my desk location was moved three times during a three-week crash apprenticeship. I was then finally trusted with project assignments and could start working.

The cramped space and lack of privacy in the office foster an admirable discipline. Common bookshelves and cabinets are designated for all the notes, disclosures, and relevant technical papers for each particular project. Since it is a public file, contributors deposit neatly prepared documents. Newcomers can access these public files quickly, instead of having to rummage through private files of previous workers on the project. Since it is customary in Japan to include many employees' names in technical papers or patents, the public files are relatively safe from opportunists within the company who may want to confiscate information.

I soon discovered that the lack of privacy at work also extended to personal time off the job. Social life, hobbies, sports, and travel are all considered corporate-related. To join an archery club, rather than turn to the Yellow Pages of a phone book, an employee would apply through the company sports coordinator. The same procedure was followed for activities like skiing, tennis, bicycling, and bowling. In most cases, the company even supplied the equipment and a sports outfit emblazoned with the Melco logo.

On business trips, employees are similarly immersed in the company culture. I cannot recall any business trip where I was allowed a private room in a hotel or even left to be by myself. My first business trip, about 300 miles by train to the Central Laboratory in Itami, was the most grueling. I stayed for seven days with three co-workers in a company dormitory where double occupancy was mandatory. Out of a sense of politeness the first night, my roommate and I continued talking in our room until early in the morning. Neither of us had any private time to prepare for presentations the next day about our circuit breaker work. As my roommate was nodding off to sleep, I slipped off to the bath by myself. Before long, I found him sitting in the bathwater next to me, apologizing that he had left me alone for that brief moment. He continued to follow me around for the rest of the time—virtually day and night for the entire trip—so that I would not feel "abandoned." I also recall a ski trip when the entire period of 60 hours—including, of course, the bath—was spent constantly in the company of other employees.

During a trying period in one of my projects, we were falling behind schedule because the computer program to analyze magnetic fields for our prototype device was not entirely user-friendly. Our section manager summoned what I thought were three young apprentices to help with the project. They painstakingly scrutinized the data and put in enough overtime to complete the work on schedule.

When I later asked my boss about these "apprentices," I was surprised to learn that one was an engineer of my rank, another was the author of the computer program, and the third was the author's collaborator. They were not apprentices at all, but expert consultants. They were what I call ghost workers—employees who often turn up when there is an engineering problem and, once the problem is solved, vanish.

On the other hand, the elite core of nurtured workers known as lifetime employees make up about one-third of the company's work force. The rest are so-called ghost workers, who come chiefly from subcontractor companies that also supply full-time professionals. These small outfits differ from U.S. subcontractors by being heavily allied to only one company.

The specialized ghost workers appear to play an important role in Japanese companies. They form a pool of expertise at a subcontractor level that companies can use freely. They also train generalists at a company to better serve the corporation in the future. And they make lifetime employment possible for the select employees, since layoffs or firings are generally done at the subcontractor level. Others who face attrition are female workers—who even as engineers are not fully considered lifetime employees—and certain people who officially retire at the mandatory age of 55 but who are retained as workers or as heads of subsidiaries, suppliers, or subcontractors. Since these employees are generally not entitled to company pension or medical benefits, there are virtually no severance costs to the company when they are laid off.

Other sources of ghost workers in a section are lifetime engineers who rotate on apprenticeships from, say, design to manufacturing. Throughout my assignment in the engineering and design departments, temporary members would join my project from either the patent, manufacturing, or marketing departments. Such apprentices become familiar with a project quickly. They not only become a backup team but also make it easier to "sell" the project outside the section to other groups in the company.

I became a ghost worker in the middle of my year at Melco, when I interned in the patent office. My project team was at the point of filing patent applications on the circuit breaker accessories we had designed and built. Since I was not familiar with the Japanese approach to patent disclosures, I asked my engineering manager for advice. He let me work as an apprentice for about two weeks in the patent section, where I learned the general procedures, costs, and success rates in filing for patents. A patent application usually contains two dates: the one when the idea is first disclosed to the company— its official discovery date—and the date when the idea is filed with the national patent office.

At first I thought that my patent apprenticeship was a special privilege because of my exchange engineer status. I soon noticed, however, that many of my engineering division colleagues rotated through the patent section for training when they themselves were involved in filing for patents.

Meetings without minutes or memos

The engineers at Melco engage in as many daily meetings as engineers at Westinghouse. Most meetings, however, are informal, with no advance agenda or follow-up minutes. In any given week there may be several morning meetings within a section. Generally there is no clear agenda, and no minutes are kept or reports issued; the gathering is meant only to inform, as opposed to a contract review. The lack of memorandums and reports—items found in abundance in U.S. offices—can be attributed in part to the inability of the office printing machines to handle the more than 2,000 Japanese characters necessary for everyday communication. Because there are no simple typewriters or high-speed word processors for the Japanese language, paperwork is routinely done in handwriting.

The informal communications and the minimal paperwork left me somewhat confused during my early months as a new member at Melco. The appearances and disappearances of the ghost workers, for example, occurred without accompanying memoranda. Coordination of project tasks was also handled by word of mouth rather than documentation. In time it became clear to me that my engineer colleagues were much less concerned with deadlines, schedules, and productivity than I had originally anticipated. The emphasis seemed more on quality of work or on doing things right the first time. But through the informal communications and frequent discussions among coworkers and supervisors—as opposed to formal contract reviews—my project work benefited from constant fine-tuning. In our design work, I also noticed that in addition to input from marketing people, our team was in touch with the manufacturing engineers from the start.

In the first week, when I began my project to design an electronic circuit breaker controller, I tried to identify experts to give advice in specific areas like electronic circuit design, magnetics, materials, and cost analysis. My four closest co-workers, I soon discovered, had all received recent training and were therefore semiexperts in these areas before they joined the project. They became my sources of advice.

Because of my lack of manufacturing experience, I was sent down to the shop floor for over a week, where the manufacturing engineers showed me a device like the one I was to design. They let me see the tools and machinery for making the product and pointed out cost trade-offs in terms of materials, labor, and the volume produced in a year. With this introduction, I stayed in touch with the manufacturing staff and received helpful comments throughout the design process. This, of course, also happens in U.S. companies, although the flow of information may not be as spontaneous. I also sensed that the status of design and manufacturing engineers was about equal in Japan and the relationship was never antagonistic.

In my group, design engineers would double as marketing experts, working out new specifications with customers and attending market forecast meetings between stints on the design bench. Because of my overseas affiliation, I was understandably not given the expanded role of a marketer.

In the Japanese approach to product development I did not see distinct groups of professionals working on the various aspects, such as R&D on proof of principle, marketing to generate specifications, design to produce working prototypes, and manufacturing to achieve economical production. Instead my manager encouraged me to spend two or three weeks away from my immediate job to delve into other areas. The project was often delayed, but I became a better generalist with time. The successful integration of manufacturing, design, and marketing at Melco can be attributed in part to freedom for the professionals to roam in and out of different disciplines informally. Again, these personnel rotations and transfers occurred without paperwork and written requisitions.

Paralleling the absence of job descriptions and clearly defined boundaries, my Melco colleagues appeared to make little distinction between office and nonoffice hours. Since most of the meetings were informal, there was no fixed time for a meeting to adjourn. Discussions often extended into the evening. Engineers with dissenting views or new ideas often waited to have their say outside the normal working hours. The discussion might be accompanied by dining and drinking to avoid the atmosphere of confrontation often associated with a formal meeting during office hours.

Impromptu discussions and late night or weekend business

conferences are common in Japan. Employees believe their time belongs more to the company than to themselves. I found this informal communication to be somewhat time consuming. It helps, however, to counteract the high degree of bureaucracy and hierarchy in the work place.

Outside the work place, I often was thwarted by a similar lack of written, clear guidelines. Yet there seems to be a paradoxical uniformity in most things the Japanese do. Their food preparation, clothing styles, and certain mannerisms are rigidly standardized. Restauranteurs prepare basic Japanese dishes with a uniformity comparable to that of a U.S. hamburger chain. No matter what the thermometer says, only after June 1 is it considered appropriate for students and even some corporate engineers to don their summer wardrobes.

As for standard mannerisms, any tourist who has visited Japan surely remembers the habitual *"Irrashaimase"* acknowledgement of department store salespeople. Similarly, engineers presenting technical papers invariably begin with a standardized introductory phrase. Unlike the practice in the United States, varying the standard opening is simply not done.

All of these cultural nuances are second nature for the Japanese; no book of rules is necessary. A visitor learns these codes soon enough; my main struggle was my inability to read Japanese well. Speaking the language, on the other hand, came more naturally after the four months of intensive training that I received from Westinghouse prior to my arrival in Japan. After six months of living in Japan, I could speak Japanese much more fluently.

Indoctrinating engineers: like a college campus

Like other Japanese companies, Melco does not emphasize personal interviews when it hires engineers. Instead it hires some 700 graduates in blocks from select schools every year. These graduates do not apply to the company for a specific job; they begin with a year-long indoctrination period in which they attend company lectures and do apprentice work on the shop floor. They sleep and eat in Melco dormitories, where friendships developed easily, much as they do on college campuses. During this first year at Melco, in fact, they are called freshmen.

After indoctrination, the new engineers make bids for particular areas and are assigned, upon mutual agreement, to a division. After five years the young engineer may request a transfer, called a rotation, or the company may initiate the transfer. Once he enters a division, a seniority system usually keeps him on the technical ladder for as long as 15 years. During this period he is moved among the groups in design, manufacturing, sales, and patents to acquire broad experience. There is no "fast track"—each engineer is trained to be a generalist.

An engineer's salary, including twice-a-year bonuses and 60 to 100 hours of overtime a month at 1.3 times base pay, is about 40 to 60 percent that of a similar U.S. engineer. The cost of living in Japan is even higher than in the United States.

Most of an engineer's waking hours outside the office—social time, sports, music, and even political campaigning for company-backed candidates—are spent with other employees of the company. My colleagues' wives were not included in any of these activities. On the rare occasions when Japanese entertain dinner guests, wives still do not sit with the guests, talk, or join in the meal. There was little home life for my Japanese engineer friends.

In one seven day period, one of my married co-workers in the design and engineering section spent nearly 70 hours in the office. I estimated that he spent about half that time doing design work at his drafting table and poring through catalogs, standards, and patents in support of his design. He spent half an hour every morning flipping through a stack of recent disclosures published by the Japanese Patent Office—an exercise that raised his competitive spirit whenever he came across an important disclosure from a rival company. He spent one to three hours that week with customers and about 30 hours in informal meetings on the job, showing drawings of prototype devices to manufacturing engineers and marketing people.

Two evenings that week he went out to dinner with his work group and sang to recorded music in a bar—a common entertainment for groups of all-male patrons, who chant to popular songs while downing beers. These two evenings of entertainment consumed eight hours. He played in the department's badminton match on another evening. Needless to say, my colleague's wife and two children saw little of him that week.

The following week, he was asked to go door-to-door campaigning in his neighborhood for a company-backed candidate in the local election. Such campaigning was more or less mandatory although it would not be done for the national elections.

Company committed to employee training

Long hours of employee dedication like this are matched by company commitment to employee training. The training at Melco does not end after an entering crop of graduates complete their year-long initiation at the company's quasi-university. Career-long training courses for engineers are prepared within the company, with the exception of some offered by such technical societies as the Japan Union of Scientists and Engineers and the Japan Institute of Standards. The training makes use of lecture courses, special assignments, and internships at other Melco plants.

Blue-collar workers in the division also receive planned training. Toward the end of my year at the company, when several prototype devices for my project were constructed, I had an opportunity to visit the model shop and testing laboratory to evaluate the prototypes. I noticed that many of these shop-floor workers and technicians were familiar with basic statistics and could present data from experiments in concise reports—a rarity in the United States. A foreman told me that time was also set aside for them to experiment with new instruments and machinery. There were many display charts at different workstations listing simple statistics on product yield, failure rates, and time-sequence comparisons.

For the testing areas and model shops, the company bought many new foreign instruments, such as an advanced welding machine, to be set aside for hands-on experimentation. One foreman in the testing laboratories explained to me that new equipment, even if not directly used in production, introduced novelty into the work routine. More important, he said, the workers get self-training and are more open to ways to improve manufacturing.

While the capital equipment fund was generous for production personnel, the offices for white collar workers were stark. Following the tradition of community property, personal computers, terminals, telephones, and photocopying machines were shared by hordes of office people. As a senior engineer, I shared a telephone for calls within the plant with four other engineers and an outside line with about 15 engineers. All day long, as calls were redirected from busy phones to unengaged ones in the large office of 80 people, beckoning shouts and messages could be heard in the office hall along with the incessant ringing.

There were, however, enough personal computers available for engineers in my department, perhaps because Melco sought our comments on its newly introduced computers. Word processors in Japanese were still in their infancy in 1983 because of the sheer numbers of characters and homonyms in the language.

In line with the Spartan ambiance, office furniture was held to a minimum. Simplicity and cleanliness were major themes at Melco. In my division we were ordered to clean up and clear the desk before going home every day. And in my visits to other Melco sites, I always found the shop floors immaculate and efficiently arranged.

The initiation of Koji

I noticed during my stay in Japan that Melco and Westinghouse engineers designing similar products carried more or less the same work load and responsibilities. The main contrast seems to be during the engineering training phase in Japan, when teamwork is encouraged. Engineers entering the company undergo two orientation phases: ''clan rites'' to initiate them into the corporate culture, and generalist training, when they learn a breadth of disciplines through rotating apprenticeships. The training of one electrical engineer colleague, whom I will call Koji, is perhaps typical.

Koji joined Melco when he was 23. While living in the company's dormitory, he worked in various corporate sections and was introduced to the nuances of each. Beyond the technical skills, he established personal contacts through communal living and joint work projects. Day and night, he was a ''Mitsubishi man'' surrounded by colleagues.

Because promotion is slow among lifelong employees and is based on seniority, Koji learned early that he would not be considered for his first managerial position until he was 38—and then only if he had attained the chief engineer's rank by the age of 33. (Chief engineers are roughly equivalent to a project leader or senior engineer in the United States.) With planned promotion like this, nearly every employee knows where he stands in a succession chart.

Koji, who was acknowledged to be one of the golden boys of his entering class, began his networking in the dormitory to muster support among his colleagues for the promotions that lay years ahead.

Periodically the new engineer took part in a formal ''family training'' program in his department. This program was comparable to the manufacturing group's quality circle, where all ranks participated in making and implementing suggestions designed to improve working conditions and products. In the engineering group, the discussions tended to stress communications techniques and accountability up and down the hierarchy, from junior engineers to the division manager.

At one session we heard an oral report from Koji on what appeared to be an ongoing accountability problem: Who was responsible for the final versions of mechanical drawings that were subsequently submitted for prototype production? Many engineers worked on each drawing, but only two signatures appeared at the bottom of the completed copy. Should the two engineers be held accountable for the drawing? What if there were major mistakes—were they to blame? Koji then moderated a panel discussion among the engineers and managers about the matter.

After an hour, the formal exercise was concluded without resolution. The participants seemed to benefit, however, from the opportunity to confront one another. And Koji, as mediator, had just experienced the equivalent of a class in Management Training 101.

During the next several years, Koji was invited to many company social functions, like a ski trip with the computer section or an island boat trip with the drafting pool. At each he was asked to give a talk about himself and his work.

Koji also participated in many training courses, in which he made friends with more engineers—friendships that would be helpful in future jockeying within the company. Koji explained that most of these activities and opportunities were planned and staged by his managers.

One uniform for all

Melco engineers wear a common uniform with no indication of rank. There are, however, a series of numbers on the name tag that pinpoint the year of entry of each employee. In this way, Koji could quickly determine the seniority of fellow workers and whether they were of the same "class" as his.

Because of my short stay at Melco, my own training phase was more socializing than technical training, bringing me a continuous stream of invitations to join in social events. I went to more company sponsored sporting events and outings in this one year than in the previous five years working in the United States.

One might pity a Japanese engineer for having to sacrifice almost all of his personal time to the company. On the other hand, I can appreciate that the company is nurturing its human assets by stressing the value of cooperation. As a result, Japanese engineers are used to working together as a team.

By the time he was 33 years old—10 years after joining the company—Koji had already spent six months on the manufacturing floor of the medium sized motor division, one month in the patent section, and one year taking master's degree courses at a university in England and traveling to various electrical companies in Europe. Each year he also attended a refresher course at Melco on solid state devices, and once he taught a class of Melco freshman on the same subject. In his third year with the company, he was rotated from the Circuit Breaker Division to spend a year at Mitsubishi's Central Research Laboratories in Hamagasaki, learning to use computer programs for mechanical stress analysis.

Koji, however, remained a section electrical engineer for 10 years. He received only one promotion, at the age of 28, when he moved in rank—as all the section EEs did—from a Class 2 to a Class 1 engineer. Heavy training and slow promotion were normal in our group; there was only one section manager for the 35 employees.

From the first day on the job, the understanding is that the engineer will stay with Melco for his entire working career. On its part, the company assumes responsibility for offering generalist and specialized training to the engineers. With the working life of the average engineer as long as 40 years, Melco deems it essential to give its staff members experience in various fields before they assume policy-making decisions. In contrast to U.S. companies, which tend to encourage outside education, the Japanese approach is oriented toward in-house training. Night courses are rare in Japan.

There are several reasons why in-house training may be preferred. For one, the course work can be tailored to fit job requirements. Most company courses are, in fact, drawn up to solve practical development problems, as in the well-known Kanban system of Toyota. Melco provides room and board, a dormitory, and class sites, as well as opportunities for employees to meet with one another and to discuss work problems. This helps promote among engineers a sense of loyalty to the company. By contrast, many U.S. engineers view the completion of course work at a university on their own time as a sacrifice, and demand recognition from the company, either through promotion or pay increases.

During Koji's training assignments, ghost workers were called in to do his engineering work. In the Japanese tradition of apprenticeship, Koji was obliged to train his colleagues when he completed a course.

In the United States, an engineer usually does not know whether his private training represents a good match with the immediate needs of the company; a postcourse presentation to peers at work is virtually unheard of.

Job rotation for trainees

With promotions being slow, Koji's training continues to this day in the form of job rotation. Many U.S. engineers would resist a lateral job transfer or rotation into other fields of work at different company sites unless there was the promise of promotion. Lateral transfer is routine in Japan, where a research engineer might undergo a tour of duty in manufacturing and vice versa. As a result, Koji, like other Japanese engineers, identifies himself less with his field of expertise or job title than with the company. He is foremost a Mitsubishi man, not an electrical engineer.

Koji has two mentors. One is his section's chief engineer, who graduated from the same university and is in line to be the next section manager. The other is a general manager in the motor division, whom Koji met while training during his freshman year. The manager took a liking to him and assumed the status of "outside" mentor, who checked periodically on Koji's career progress. He recommended vital training courses and looked out for career opportunities that Koji or his chief engineer mentor might otherwise have missed.

In general, the company assigns a "big brother" engineer during the first year of training. This mentor is often not a member of the trainee's immediate group, so that the group's section head will not lose face later on when a good man leaves his outfit. Jobs in sales and marketing especially make use of outside mentors.

The day I arrived at Melco, one of my co-workers was assigned to be my big brother. He well understood that an exchange engineer's purpose was not only to perform the assigned project but also to learn the Japanese organization and culture. In contrast to earlier U.S. visitors, who arrived at Melco as licensing experts and were regarded as teachers, an exchange engineer was more of a peer. My big brother had to explain this new peer concept to the other engineers to avoid misconceptions.

Slow going for female engineers

Throughout my stay at Melco, I had no opportunity to work with a female engineer, for a simple reason: in my division of some 2,000 engineers, not one was a woman. During visits to other locations, however, I met women ranked as engineers, who discussed their work willingly.

The most common job for women in Japan is that of the "office lady," as she is called. The job duties include answering phones, making and filing drawings, serving tea, making copies, and setting out lunch boxes for the office staff. In my engineering office, there was no secretary. Secretarial work was done by the engineers and by the office ladies on a temporary basis.

Melco first opened its doors to female engineers in 1981, largely to enhance the company's software capabilities. There was then, as now, a great shortage of software engineers. One such female engineer I met had started with the company in 1981 and earned as much in base salary as her male counterparts—well above the salary of office ladies. Nevertheless she and her female engineering peers were still called on to do such side jobs as serving tea and setting out lunch boxes.

Career competition with her male counterparts was difficult, this female engineer said. Both national law and corporate policy limit the amount of overtime for women. At Melco men can work four times longer than women on paid overtime. In practice, engineers of both sexes often work far more hours than the company pays them for, however.

Myths behind group consensus

During my stay at Melco I participated in many decision-making meetings. Because we were working on a new product line, my project team was constantly seeking the latest R&D and marketing news to help plan the direction of our work. But, contrary to what I had read before going to Japan, I did not see a particularly strong drive toward consensus decision making.

Some U.S. writers have contended that company decisions in Japan are made only after everyone involved has expressed an opinion and there is general agreement that the decision is a good one. Although this is not completely off-target, what I observed seemed to be simply an outcropping of a clan system where responsibility for a unilateral decision is spread evenly to make ''accomplices'' of group members. The ritual of passing proposals around for everyone's stamp of approval is one example.

To attempt to gain backers for a proposal, my Japanese colleagues and their managers preferred a personal lobbying approach rather than an open, face-to-face meeting of the people at odds. Most of the negotiations and horse trading in my immediate group at Melco always took place before a crucial meeting was held. This lobbying and ironing out of differences usually took place after office hours.

During my year at the company I heard a rumor that the division was toying with the idea of getting a computer-aided design (CAD) package. Commercially available packages could be readily purchased, but the division would then have to rely on the vendor for years to come for technical support. The division staff quietly performed a cost–benefit analysis and ''proposed'' that the CAD be developed in-house. As the rumor of that decision reached the engineering department, an older group of engineers grumbled among themselves, saying that too many engineers would be committed for a long-term effort to reinvent a commercially available product.

Nevertheless, the decision appeared to have been made unilaterally by the management. A proposal outlining the decision was soon written by the division staff and circulated in the department. The proposal's front page was soon covered with employees' names stamped by means of their seals, along with the dates when the page was stamped. Some stamped names were upright, others were not, possibly to denote the degree of support. My "big brother" told me that stamping one's name on a document did not necessarily convey endorsement in the way that a signature might in the United States. The seal signified only that the person had seen the document—in the same way that when a Japanese means "I heard you," he utters a word akin to "yes." As the proposal collected more and more stamps, even from those only remotely affected by the decision, the remaining employees found it easier to stamp their names on the document, knowing they had lots of company.

To foreigners, the placing of many employee-seal stamps on a proposal—sometimes as many as 50 names—appears to indicate at first glance that consensus approval plays a primary role in Japanese decision-making. However, to make an informed decision, a Japanese chief sometimes has to call on young, low-seniority key players. Not to embarrass their more senior colleagues in the status conscious hierarchy, the chief brings many others in the division into the decision-making process. Such an inefficient stamping ceremony could easily be hailed by outside observers as the Japanese consensus culture.

In open meetings my Japanese colleagues were characteristically uncritical, reluctant to negate anything—a characteristic of the language itself. They appeared much less flamboyant than my U.S. colleagues were in free-form discussions such as brain-storming sessions. When I once called a meeting in Japan to solicit comments on my design work, the lack of critical comments disappointed me. I later received dinner invitations from many of those at the meeting. Only during the dinners did I hear criticism, in varying degrees of directness.

By the tenth month of my year's stay, I prepared a technical paper on behalf of my Melco co-workers for an annual conference in Osaka sponsored by the Japanese Institute of Electrical Engineers. Knowing that competing company engineers and scientists would be present, my co-authors coached me thoroughly on the questions I could expect from the floor. When the time came, everyone at the conference seemed extremely polite during the discussion period, in contrast to the very frank and sometimes antagonistic remarks at some U.S. technical conferences.

Our paper presented a magnetostatic field analysis of a certain electromechanical device. One conferee commented that the analysis was probably very costly in comparison with the device under study—could we justify the effort?

My polite answer was that the conference session was of a technical nature. The economic aspects of the device, I suggested, should be addressed in other appropriate sessions. It struck me later that I had finally adapted, after many long months, to the Japanese way.

Reprinted from *IEEE Spectrum*, Sept. 1985, pp. 72–81.

21/JAPAN'S INTELLECTUAL CHALLENGE

Lawrence P. Grayson

Part I—The strategy

On August 15, 1945, Japan lost a major war that had devastated the country, both physically and sociologically. Japan—the country that isolated itself for over 200 years from Western progress, only to be awakened by Perry's arrival in 1853; that as late as 1868 had a feudal system under which merchants and businessmen were in the lowest class; that had built a manufacturing industry by the early 1900s, but then focused it on military needs; that at the end of World War II was in ruins with its industry at a standstill, its urban population reduced by half, and 40 percent of the aggregate area of its cities destroyed—had to rebuild itself and do so under new conditions. The Allied Forces demilitarized industry, prohibited monopolies, dissolved family controlled industrial combines known as *zaibatsu*, banned militarism and nationalism in the schools, and opened education to more people than just the privileged minority.

In less than 40 years, Japan has achieved such huge economic success that it is the second largest economy in the world, surpassed only by the United States on which it is rapidly gaining. In 1950, Japan's GNP per capita (a measure of how much each person produces and, ultimately, of how well he lives) was 6.5 percent that of the United States; by 1982, it was 75 percent and increasing about twice as fast. In 1981, the United States had a $48.5 billion deficit in world trade, while Japan had a $8.3 billion surplus; in bilateral trade of merchandise, Japan enjoyed a $17.9 billion advantage [1]. During the 12 years from 1970 to 1982, the average hourly compensation of Japanese workers in manufacturing grew almost sixfold, while U.S. workers' compensation increased less than three times. This is a direct reflection of increases in industrial productivity, which over the same period grew 108 percent in Japan, but only 31 percent in the United States [2].

Japan is in a strong position to continue its progress. Its GNP has increased from $10 billion in 1950 to over $1 trillion today. It has an extensive base of modern, capital-intensive industries and is building new industrial capacity at a more rapid rate than is the United States. Further, it is well positioned to compete for a larger share of the world market, with an extensive structure for exports, a large trade surplus and an extremely strong currency. Julian Gresser, president of the East Asian Consulting Group, in describing the industrial competition faced by the United States, has stated metaphorically, "The technological battle with the Japanese is really an industrial equivalent to the East-West arms race [3]."

There are many reasons why the Japanese have made such tremendous economic progress. They have a clear economic policy that has guided government actions. They have the highest rate of personal savings and capital formation among the major industrialized countries. Because equity financing in Japan is through banks, rather than public stock offerings, corporate management can focus on long-term growth rather than short-term profits. (Of the world's 30 largest banks, 11 are Japanese.) They spend little on defense. Government and industry work closely, and trade practices protect the home market. A strong work ethic and company loyalty have been fostered by the almost paternalistic attitudes of industry. These are the most commonly cited reasons.

Another key factor, however, is Japan's strong commitment to education, particularly engineering education. Since the latter half of the nineteenth century, when modernization began, education has been prominent in the policies of successive governments [4]. Japan has stressed the expansion of scientific and technical education at all levels, both to provide the engineers and technicians needed by industry for growth and technical development, and to produce the technically literate population required to facilitate the transfer and adoption of technology on which the nation's industrialization has depended. As a result, Japan has been able to assimilate with ease new ideas and techniques from other countries and adapt them to fit Japanese needs.

In the past 30 years Japan has supported and several times revised education as part of its plans for economic development. During this period industry demands have significantly affected educational policies. This influence is especially important, since Japan's success rests on the continued development of technology-based industries, which require people with an advanced level of knowledge and access to technical information. That education could become an instrument of economic policy stems from the close relationship between business and government in Japan. John Kenneth Galbraith describes the relationship this way:

> The industrial system, in fact, is inextricably associated with the state. In notable respects the mature corporation is an arm of the state. And the state, in important matters, is an instrument of the industrial system [5].

Development of educational policy

The period under the American occupation, from the end of the war in 1945 until mid-1951, was devoted first to demilitarization, then to reform and democratization of Japanese society, including education, and then to the beginnings of economic rehabilitation. Education was restructured after American models, as the occupation forces tried to establish educational goals similar to those in the United States. The number of educational institutions, particularly colleges and universities, rose significantly, and advanced education was made more widely available.

After independence in May 1951, the process of adapting American ways to Japanese needs and culture began. A major concern addressed immediately was the development of education as a means of promoting economic growth. One month after independence, the National Diet passed the Industrial Education Promotion Law, which recognized that "industrial education is the basis of the development of the industry and economy of our country [6]." Private industry stressed the same theme, as Nikkeiren (the Federation of Employer Organizations), an influential group representing major economic organizations in Japan, called for revisions in education to make it more directly relevant to industry's needs.

In a series of reports issued between 1952 and 1959, Nikkeiren recommended the promotion of scientific and vocational education in elementary and middle schools, the improvement of technical training for working youths, the creation of five–year technical colleges that would combine vocational high schools with two years of college, and more training in college-level engineering and science. Technical colleges were seen as essential for training the middle-level personnel required by industry. The organization's concern went beyond domestic development and recognized that continued industrial and economic growth would depend on competing in the international marketplace:

> If Japan does not develop systematic training of engineers and experts in the midst of its epoch-making economic growth in order to attain further progress in industrial technology, our industrial

technology will lag day by day far behind the international level. This will result in our failure in competition with other nations [7].

The government responded to the demands of industry. In 1957, the Economic Planning Agency, established to coordinate policies among government departments with regard to long-range economic planning, estimated that by 1962, 27,500 science and technology graduates would be needed annually. Later in 1957, the Ministry of Education issued a five-year plan for expanding the number of science and technology graduates by 8,000 [8], and announced major curriculum revisions for elementary, lower and upper secondary schools.

The coupling of economic and education policy continued, as the government adopted in 1960 the National Income Doubling Plan, which set the goal for the decade. The Economic Council of the Economic Planning Agency, which drafted the plan, underscored the importance of education when it stated that "economic competition among nations is a technical competition, and technical competition has become an educational competition [9]." The Economic Council recommended immediate action to educate a large number of scientists and engineers of high quality so that the economic plan would not be handicapped by the lack of human resources.

The Ministry of Education, acting in harmony with the economic goals, developed a plan in 1961 to raise the number of places in science and technology faculties from the then 28,000 to 44,000 within seven years. As part of this enlargement, 19 new five-year technical colleges, comprising three years of upper secondary school and two years of college, were established in 1962 [10].

As the educational system expanded, planners realized that education should fulfill a broad range of purposes. In 1963, the Economic Council proposed a diversified but meritocratic system of education. Its report, which made recommendations for all levels of education, including education in industry, again called for expanding programs in science and technology, which strongly affected educational policy planners.

The ten year goal of doubling the national income was achieved in seven years, and Japan experienced unprecedented prosperity. Although several readjusted economic plans were developed in the 1960s, the educational plans remained the same.

One reason that education policy and its implementation could be so responsive to science and technology policy is that the Ministry of Education was organized in the 1960s to receive advice from the scientific community. Reporting to the Ministry in 1966 were a Higher Education and Science Bureau, a Science Encouragement

Committee, and a Science Education and Vocation Education Council. In addition, the Ministry had direct responsibility for the budgets of the national universities, colleges and junior colleges, their attached research institutes, and a National Training Institute for Engineering Teachers [11], giving it a great deal of influence with those institutions.

In 1970 the Economic Council published its new economic and social development plan, which again broadened the purposes of education. First priority remained the improvement of science and technical education. To cope with the increasing internationalization of industrial activities, the report suggested teaching the knowledge, skills and other necessary qualities needed to develop international cooperation [12], a theme that continues to receive increased attention.

Emphasis on engineering

Since the end of World War II, Japan's economic strategy has targeted certain industries for growth and assistance to make them globally competitive, with each newly identified industry being more technologically advanced (and potentially adding higher value to the economy) than previous ones. The Japanese have become major exporters, first of textiles, then of steel, automobiles and consumer electronics, and now of semiconductors. They are the world industrial leaders in robotics and optical electronics, and are making gains in computers, telecommunications, and genetic engineering, industries that are nationally important for the 1980s.

In semiconductor memory chips, Japan's progress has been rapid. In 1970, when the 1K RAM chip (a random-access memory chip that can store up to 1,024 bits of information) was standard, the United States totally dominated the industry with almost 100 percent of the world market. By 1974, when the 4K RAM was the state-of-the-art, Japanese manufacturers, as noted in Fig. 21-1, had about 5 percent of the world market. With the introduction of the 16K RAM, the Japanese gained about 40 percent of the world market by 1978, and with the 64K RAM, Japan's share rose to 70 percent of the market by the end of 1981 [13]. In 1982, the United States used 31 million 64K RAMs, but its industry produced only 20 million chips; the difference was provided by Japanese manufacturers. U.S. imports of semiconductors from Japan exceeded its exports to Japan for the first time in the third quarter of 1976, and the gap has grown ever since. At the beginning of 1982, bilateral sales of semiconductors between the United States and Japan amounted to almost $750 million. U.S.

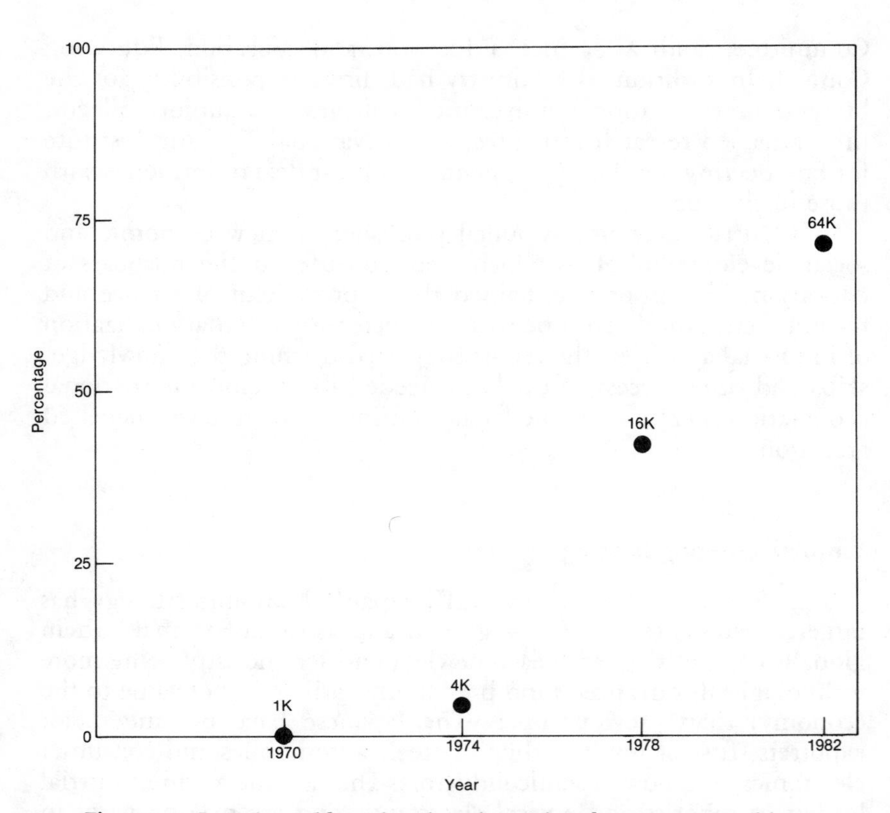

Fig. 21-1: *Japan's world market share in semiconductor memory chips.*

revenues for exports to Japan totalled only 25 percent of this amount, while its payments for Japanese imports accounted for the remainder [14]. That is significant, not only because of the amount of the revenues, which have been projected to reach $1 billion in sales in 1984 and $2 billion in 1986, but because the 64K RAM, and the future 256K and 1M chips, are essential to successful competition in the computer, telecommunications, and other high technology industries, as well as in a wide variety of consumer products for personal, entertainment and household uses.

To provide the necessary technical manpower to achieve their nation's goals, economic and education policymakers in Japan have supported increases in the number of engineering schools, engineering enrollments and faculty members. In the 27 years from 1955 to 1982, as shown in Fig. 21-2, the number of bachelor's degrees in engineering awarded in Japan rose from 9,613 to 73,593. During the same period, the number of bachelor's degrees awarded by United

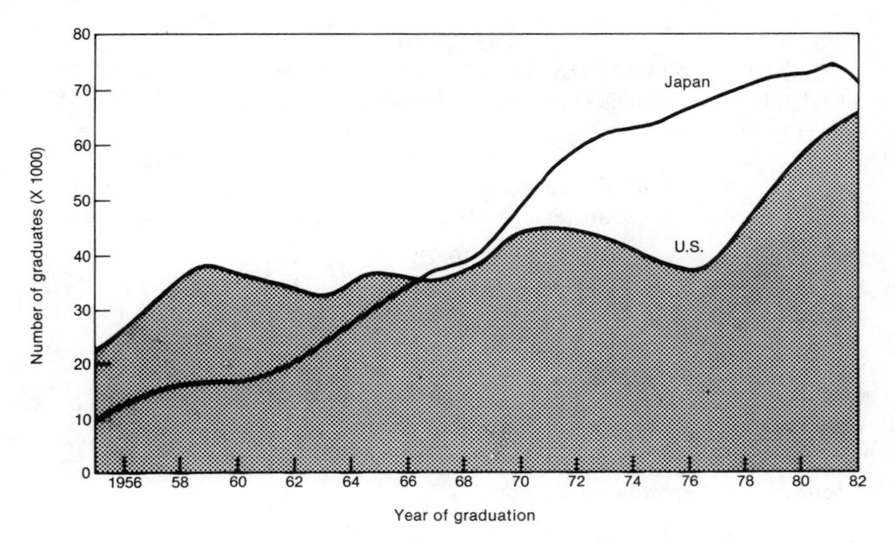

Fig. 21-2: *Bachelor's degrees in engineering in the U.S. and Japan.*

States engineering schools increased from 22,589 to 66,990. Japan graduated more engineers at the baccalaureate level than did the United States for the first time in 1967, and the United States has not equalled Japan's total since. The 10 percent difference in number of 1982 graduates is more dramatic when one realizes that Japan has approximately half the population of the United States (115 million people to America's 220 million), and that virtually all engineering degrees in Japan are granted to Japanese citizens; in the United States 8.1 percent of engineering baccalaureates are awarded to foreign nationals who must leave the United States upon completing their studies. Excluding foreign nationals, on a per capita basis Japan currently is graduating 2.3 times as many engineers at the bachelor's level as is the United States.

Educational development and economic growth have been linked reciprocally in Japan. The increased quantity and quality of educated people have provided the manpower necessary for industrial development. In turn, economic growth has enabled the country to afford, and created the necessity for, the great expansion in the number of upper secondary school and college graduates.

Although the relationship between a country's level of technical manpower and its industrial productivity is more complex than simple cause-and-effect, it is interesting to compare the average increase in bachelor's degrees in engineering conferred with the rate of productivity growth, measured as output per hour in manufacturing, in the United States and Japan. As shown in Table 21-1, changes

TABLE 21-1
AVERAGE ANNUAL RATES OF INCREASE IN BACHELOR'S DEGREES IN ENGINEERING CONFERRED AND IN MANUFACTURING PRODUCTIVITY

	Japan		United States	
	Increase in Graduates	*Productivity Growth*	*Increase in Graduates*	*Productivity Growth*
1950–1955	n/a	10.9%	−15.6%	2.7%
1955–1960	11.2%	7.8	10.9	1.2
1960–1965	13.1	8.5	−0.7	4.4
1965–1970	10.0	13.2	4.0	1.2
1970–1975	6.2	6.8	−3.0	3.4
1975–1980	2.3	8.0	8.7	1.7
1980–1982	0.2	0.9	7.4	0.9

Source: U.S. Department of Labor

in productivity growth since 1955 have almost always been parallel to—but lagging by several years—changes in the number of graduates in both countries [15]. In contrast to Japan, which has made a national effort to assure that its industrial growth would not be slowed by a shortage of qualified people, a recent report of the Office of Technology Assessment has concluded, "In general, the United States appears to have more low-skill manpower and less high-skill manpower than an industrial economy of the 1990s will require [16]."

Engineering education, however, is only one aspect of Japan's commitment to developing human resources and preparing its people and industries for a high technology-based society [17]. The nation has a strong elementary and secondary education system whose graduates are highly knowledgeable and supportive of science and technology; a higher education system that produces a large pool of engineering and technical manpower; and extensive in-company education and training to make employees most effective in meeting both short-term and long-term goals of the corporations. This structure is supported by a culture and society that places a high premium on achievement in education, an emphasis that permeates Japanese society.* It is reflected at all levels of students from the preschooler to the employed adult, in the industrial policies for manpower development and the cultural concern for academic

* For a comparison of U.S. with Japanese primary and secondary education, see Grayson, L. P., "Leadership or Stagnation: A Role for Technology in Mathematics, Science and Engineering Education," *Eng. Ed.*, Feb. 1983, pp. 356–366.

achievement, in the emotional support of the student by his family with its "*kyoikumama*" (i.e., "education mama"*) and by the hundreds of thousands of *ronin*—students who, having failed to gain admittance into preferred universities, will continue to study and reapply the following year or even for several years. Educational achievement determines one's opportunities through life and is a central factor in national development.

Industry's approach

Industry plays an important part in the overall educational process; lifelong education and on-the-job training are integral parts of the policies of Japanese industry. Most large companies, including Nippon Electric, Matsushita, Hitachi, Fujitsu, Sony and others, run company schools and educational institutions for all employees—blue-collar, clerical, engineers and other professionals, and managers.

In the early twentieth century, as Japan was beginning its industrialization, there was a severe shortage of qualified personnel and a high turnover in most types of labor. To retain a skilled work force, large companies evolved a system of lifetime employment. Salaries were determined by seniority rather than performance, many benefits including health services and recreational facilities were provided, and many activities of high motivational value were introduced to help employees identify their future with the company's. As lifetime employment evolved, Japanese companies—as a strategy for the future—became willing to invest substantially in developing their employees.

Education is an important aspect of employee development. Typically, newly hired engineers will spend from a few weeks to six months in classroom and job site training under the supervision and guidance of more experienced engineers. Only then will they begin work, under the continued supervision of senior staff. During this period and through their careers, they can engage in in-plant seminars, professional conferences and on-the-job training. When they are in junior and senior management positions, they most probably will attend company conducted schools, either for management or technical training. During this period they also might be chosen to participate in a survey tour abroad to learn firsthand about certain technologies or management techniques in other countries, or

* The Japanese mother typically plays a strong role in directing the education of her children, making any sacrifice to ensure they get into the best schools.

for a company sponsored scholarship to study abroad. Most of this advanced study seems to be done in the United States. It is common among Japanese executives today, both in industry and in government, to have studied, either as students or visiting scholars, in the United States.

A substantial portion of the Japanese students who study in the United States, particularly at the graduate level, are likely to be employees of Japanese industry or government. In Japan, the typical age for a baccalaureate graduate is 22, and only 4 to 5 percent of the new graduates proceed immediately to graduate school. Most of the remainder enter employment. Yet, 43 percent of the Japanese students studying in the United States in 1979–80 were between 23 and 30 years of age, with half of them over 25. This age distribution fits the pattern of many large companies and government ministries, which typically send some of their brightest young employees to the United States for advanced education.

Under the lifetime employment system, it is economically sound for large organizations to provide some young employees with extended periods of training in areas related to their future responsibilities. Typically, after they are employed for a few years and have learned the needs of their organization, a few of the brightest and those with potential for technical or managerial leadership, are sent abroad for several years to study at the best foreign universities, with their salaries and expenses paid by the organization. Many study in fields related to technology, finance and, particularly, business. The Japanese consider American business training the finest in the world. They wish to develop a knowledge of American business affairs and make contacts that will help future leaders throughout their careers [18].

Studying in the United States

The United States has been the world's economic and technological leader, as well as a close ally of Japan, since the end of World War II. Japan, naturally interested in America and its technological developments, has sent large and consistently increasing numbers of students to study in the United States. As shown in Table 21-2, the total number of Japanese post-secondary students studying in the United States has risen from 2,168 in 1959–60 to 12,260 in 1979–80, and reached 13,610 in 1982–83 [19].

Currently, about 41 percent of these students are in baccalaureate-level programs, 32 percent in graduate schools, and the remainder divided equally between students at the associate degree level and

TABLE 21-2
DISTRIBUTION OF JAPANESE STUDENTS STUDYING IN THE UNITED STATES BY DISCIPLINE

Academic Year	*Engineering**		*Natural and Physical Sciences***		*Business and Management*		*Other*		*Total Number*
	Total	%	Total	%	Total	%	Total	%	
1959–60	216	10.0	337	15.5	318	14.7	1,297	59.8	2,168
1969–70	506	12.2	503	12.1	660	15.9	2,487	59.8	4,156
1979–80	1,018	8.3	581	4.7	2,256	18.4	8,405	68.6	12,260
1982–83	1,097	8.1	592	4.3	2,457	18.1	9,464	69.5	13,610

* Includes computer science in 1979–80
** Includes biological sciences, chemistry, geosciences, mathematics and physics.

those with non-degree status, including those intensively studying English. The result is that many Japanese students have been educated in the American approach to teaching engineering, science, management, economics and other disciplines, and also have been better prepared to deal in international matters by improving their English and learning about Western culture and way of life.

The number of Japanese students studying engineering in the United States has risen consistently for over 25 years, increasing from 146 students in 1955 to over 1,000 students in 1982. Although recent data gathered in the United States have not categorized these students by academic level, for the 12 years from 1964 to 1976, about 70 percent of the engineering students from Japan studied at the graduate level. The majority of Japanese engineering students enrolled in electrical or mechanical engineering from 1955–56 to 1970–71, as shown in Fig. 21-3, while less than 10 percent enrolled in industrial engineering. The Japanese seem more interested in learning from America the fundamental disciplines that underlie their product lines, rather than the United States approach to manufacturing and production.

In contrast, few Americans have studied engineering in Japan. In the last 20 years, not more than seven Americans were enrolled in Japanese engineering programs in any one year, and for most years there were none. Today there is not a single U.S. college-sponsored study-abroad program that allows U.S. students to study engineering in Japan [20], in spite of the fact that the United States can learn a great deal from the Japanese in manufacturing, production and management techniques.

The Japanese also have sent many students to the United States to study natural and physical sciences. From 1964 to 1973, there was an

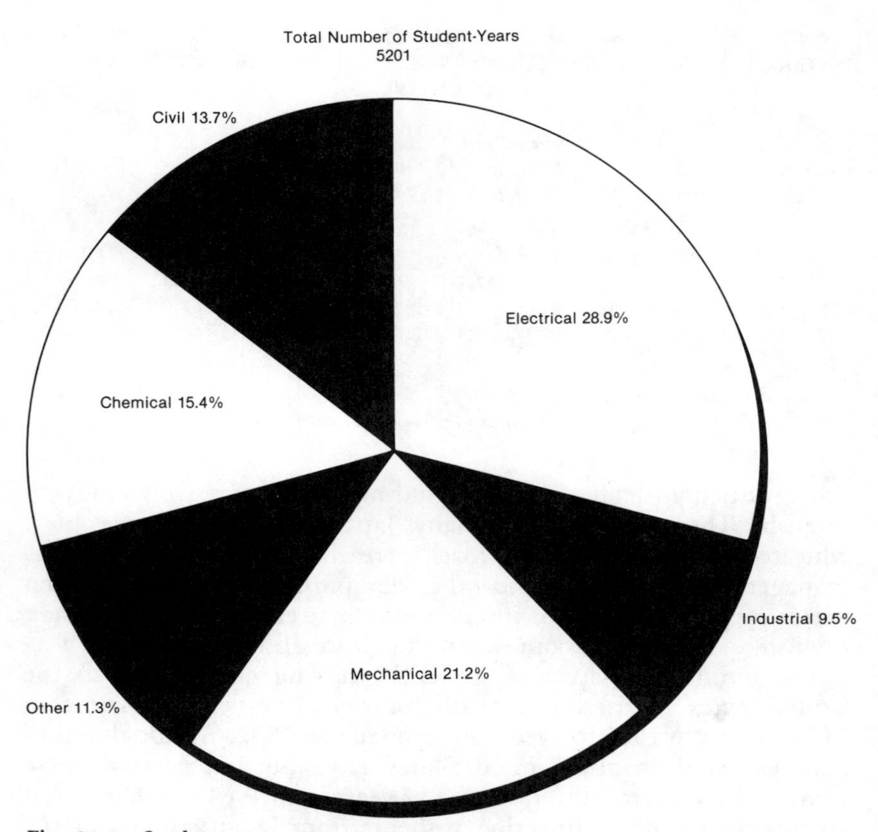

Fig. 21-3: *Student years spent by Japanese students studying engineering in the U.S. during the period 1955–56 to 1970–71 by discipline.*

average of 447 Japanese students per year studying these subjects in the United States, compared to an average of 460 Japanese students studying engineering in the United States in the same period. The United States in comparison sent an average of 12 students a year to study in the sciences in Japan during the same period. It is interesting to note that in 1973 the combined enrollments for master's and doctoral students in Japanese universities was about 5,000 students. Thus, the United States was educating about 5 to 6 percent as many Japanese graduate students in the sciences as were the Japanese universities. In engineering in 1973, about 14,000 students were enrolled in graduate engineering programs in Japanese universities, while 399 Japanese students were enrolled in United States graduate engineering programs. This lower percentage probably reflects the Japanese approach of gathering fundamental knowledge from abroad, and then developing their own methods of applying it to Japanese needs.

The Japanese, however, do not limit the people they send to the United States to enrolled students. They also include faculty members and scholars. In the 16 years from 1957 to 1973, 1,097 Japanese faculty members and scholars in engineering and 7,207 in the natural and physical sciences studied for a month or longer at U.S. universities. In the same period, 65 U.S. engineering faculty members and 317 in the sciences studied in Japan. This approach has provided the Japanese a large pool of highly educated people who return to their universities and research laboratories with first-hand experience of the state of knowledge in the United States to continue their research and educate the next generation of specialists.*

A summary perspective

Education in Japan has been an instrument used to achieve economic ends. That has been true since industrialization began over one hundred years ago, but it has been particularly used in this way since the 1950s. Government policies and private industrial goals have become intertwined, providing harmony between political, national development and private economic motives. This has enabled the state to initiate policies and take actions that directly enhance the development of industry and enlarge corporate profits. Industry in turn provides the jobs, products and services required for rapid economic growth.

As the economy expanded, industry required, demanded and received, first, large increases in the numbers of engineers being educated, then the creation of new technological colleges to produce middle-level technical personnel and, more recently, changes in curricula to reflect a broader international awareness. While the size, direction and rapidity of the educational developments are due to economic and political imperatives, the educational system that has evolved has been shaped by the culture of the country. This has produced an educational system that is in accord with the Japanese temperament, and has facilitated economic growth by satisfying the needs of industry for entry-level manpower.

* In sharp contrast, although Korea sent significantly more students to study engineering and natural and physical sciences in the U.S. during 1964–73, the number of faculty members who came to study totalled only 109 in engineering and 671 in the natural and physical sciences. Further, it appears that many of the Korean students remained in the United States, providing the U.S. with a net gain in manpower, while the Japanese students returned to Japan.

Part II—The system

The Japanese, with an intense desire for education at all levels of society and an educational system that maintains high standards, have produced what probably is the best educated population in the world. Although global assessments are difficult to make, in several international comparisons in mathematics and science conducted since 1964, no country has surpassed the Japanese in overall mastery of subject matter.

Not only is the average achievement of Japanese students very high, but the range of performance is concentrated at a higher level than elsewhere. Compared to the United States, where perhaps as much as 20 percent of the population is functionally illiterate, in Japan illiteracy is estimated to be below 1 percent. Ezra Vogel, a sociologist who has studied Japan for many years, comments on differences between U.S. and Japanese schools:

> It is commonly understood that those Japanese who attend elementary and junior high school in comfortable American suburbs will be a year or two behind their grade level in mathematics and natural sciences when they return to Japan. The same is true even for the physical education skills stressed in Japanese schools, to say nothing of Japanese and Chinese history...[21].

Vogel noted similar gaps in musical and artistic skills. Japan has a well-educated labor force receptive to learning specialized skills or being retrained at the workplace.

General education

Japanese education is composed of six years of elementary school, three years of lower-secondary, three of upper-secondary and four years of college. Although only nine years are required, over 94 percent of the students continue to upper-secondary school, and almost all of those (95%) complete that level [22]. In the United States, in comparison, only 72 percent of the 17- and 18-year-olds in 1979–80 had completed four years of high school [23].

Classes in Japan meet five-and-a-half days a week for 35 weeks or more a year. The school day, which runs from 8:30 a.m. to 3:00 p.m., is divided into six subject matter periods for five days and four periods on Saturday. Although 210 school days per year are required, most schools are open 240 to 250 days a year [24]. Compared to 180 school days in the United States, Japanese schools provide one-sixth

TABLE 21-3
TOTAL TIME PRESCRIBED FOR SUBJECTS IN JAPANESE ELEMENTARY AND LOWER-SECONDARY SCHOOLS AND AVERAGE TIME ALLOCATIONS IN TYPICAL U.S. CLASSROOMS

	% of Combined Prescribed Time in Elementary and Lower-Sec. Schools, Japan	*Annual Hrs. Prescribed for 5th Gr., Japan*	*Annual Hrs. Typically Spent in 4th Gr., Japan*	*Annual Hrs. Typically Spent in 5th Gr., U.S.*
Reading and language arts	22.2%	157.5	210	330
Mathematics	15.6	131.25	157.5	135
Science	10.2	78.75	78.75 }	51
Social Science	10.6	78.75	131.25 }	
Music and art	13.3	105.0	105 }	195
Physical education	10.5	78.75	78.75 }	
Foreign language*	3.9	—	—	—
Other non-academic subjects**	13.7	131.25	131.25	—
Total	100.0%	761.25	892.5	711

* Foreign language teaching begins in the 7th grade and constitutes 10 percent of the classroom time in the lower secondary school.
** Industrial arts, homemaking, moral education and special activities.

to one-third more classroom contact hours. Thus, on the basis of time in class alone, the average high-school graduate in Japan has the equivalent of two to four years' more schooling than the United States high-school graduate.

Further, more concepts and skills are taught, at greater levels of difficulty and at earlier ages, in Japan than in the United States [25]. As noted in Table 21-3, national school standards in Japan require that over 25 percent of the class time in the elementary and lower secondary schools be devoted to mathematics and science, subjects introduced in the first grade and continued throughout the nine years of compulsory education [26]. Elementary school students are taught such mathematical concepts as correspondence of geometric figures, and probability and statistics [27]. These are topics that in the United States are usually reserved for high school or college [28].

In upper-secondary school, the curriculum is divided into two streams. Students in the academic course of study, which is completed by about 30 percent [29], take three years of mathematics, including at least the elements of probability and statistics, vectors, differentiation and integration, and the functions and programming of computers [30]. Students in the non-academic and vocational streams also may complete three years of mathematics. They study the same subjects, although their courses have an applied orientation. Further, at more than 100 upper-secondary schools, students interested in pursuing careers in science and mathematics may elect an intensive curriculum that includes 18–20 credits of mathematics. This is in sharp contrast to the United States, where only one-third of the recent high school graduates have completed three years or more of mathematics. At each high school grade level, as shown in Fig. 21-4, more students in Japan study mathematics than do students in the United States [31].

A similar emphasis prevails in the sciences. In upper-secondary school, many students complete three natural sciences (physics, chemistry, biology, or earth science). Courses in Japan draw heavily on U.S. biology, chemistry and physics curricula, such as those of the Physical Science Study Committee (PSSC). Materials are imported,

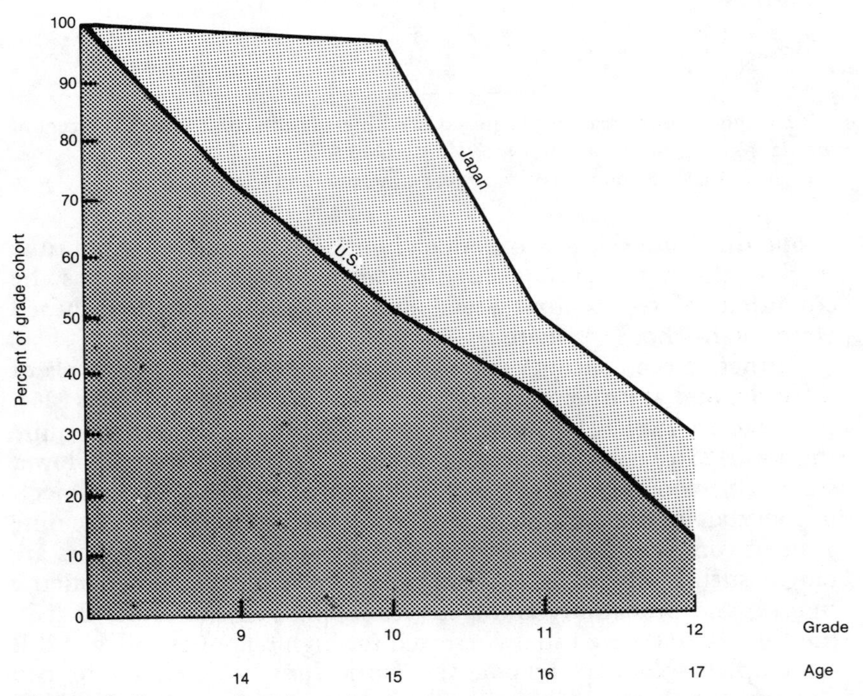

Fig. 21-4: *Percent of students in the U.S. and Japan enrolled in mathematics courses, by grade level.*

seminars held with American specialists, textbooks prepared, and laboratory equipment adapted to Japanese requirements. Newsletters, workshops and published papers promote the teaching approaches. Japanese scientists and science teachers have also developed new types of laboratory equipment, often substantially funded by the Ministry of Education [32], and, at local initiative, have established Science Education Centers in each of the 46 prefectures (or districts) throughout the country, where thousands of teachers have received in-service training in mathematics and science.

Foreign language teaching begins in the seventh grade, with about 10 percent of class time devoted to it. Although students have a choice of languages, almost all elect English. By the end of upper-secondary school, most of the college-bound students have studied six years of a foreign language.

The strong academic orientation of Japanese students can be seen from Table 21-4, which shows that they spend significantly more time doing homework and less time in non-academic pursuits than the typical U.S. student. All Japanese students, including those who will become craftsmen or production workers or pursue other vocational trades, are well educated in mathematics and science. The average factory worker in Japan is more likely than his American counterpart to be capable of discussing and implementing quality control techniques based on sampling procedures, describing in analytic terms a production problem on the shop floor, and being retrained easily for more skilled jobs that may be created by automation.

The educational system in Japan is deeply affected by an extreme reliance on examinations to determine advancement. The Joint Achievement Test, administered nationally every year, is the primary determinant for admission to college. Excellent performance significantly increases a student's chances to be accepted by a well-known university. This nine–hour series of examinations tests scholastic achievement in the Japanese language, two areas of social studies, mathematics, two sciences, and a foreign language (in 1980, 99.7 percent of the students chose English). Since in the Japanese system graduation from college is virtually guaranteed after acceptance, and many large businesses limit their hiring to graduates of certain prestigious institutions, this examination in effect determines future opportunities and success.

Teaching, as a result, is geared to preparing students for the Joint Achievement Test. As most upper-secondary schools also require subject-matter achievement tests for admission, lower-secondary schools have become examination oriented, a pattern that continues downward through the system. Many parents send their children to

TABLE 21-4
OUT-OF-SCHOOL ACTIVITIES OF U.S. AND JAPANESE HIGH SCHOOL SENIORS BY PERCENTAGE OF TOTAL, 1980

	Japan			United States		
	All Students	*Male*	*Female*	*All Students*	*Male*	*Female*
Hours Per Week on Homework:						
Less than 5	35	37	33	76	80	71
5 to 10	29	24	35	18	15	21
More than 10	36	40	32	6	5	8
At Least Once or Twice a Week:						
Reading for pleasure	29	31	27	45	39	51
Going out on a date	9	9	8	57	56	57
Just driving or riding around	14	17	9	60	65	56
Reading front page of newspaper	62	65	57	69	72	65
About Every Day:						
Visiting with friends at a gathering place	48	49	47	26	30	22
Talking with friends on a telephone	7	4	11	52	42	60
Thinking or daydreaming alone	25	22	28	50	42	57
Talking with parents about experiences	40	46	33	22	15	28

extra-study schools known as *juku*, where attendance ranges from only a few percent of students in early elementary school to 40 to 50 percent of seniors in upper-secondary school. The desire to attend the better upper-secondary schools is so great that a new class of *ronin* has been established, as students who have failed to gain admission to a preferred upper-secondary school devote another year to preparation rather than accept a school of lesser quality [34].

Although the emphasis on examinations skews teaching towards the most able students and leaves little opportunity for flexibility in subject matter, it has produced for Japan a highly effective education system, especially in mathematics and science. In 1970, Japanese youth in both the 10- and 14-year-old age groups scored first among 19 countries in each of a series of international science tests in biology, earth science and chemistry, and physics. The United States ranked 15th overall, surpassing only Chile, India, Iran and Thailand [35]. In the first International Project for the Evaluation of Educational Achievement, conducted in 1964, which compared the abilities of students from 12 industrialized nations, the Japanese 13-year-olds ranked first in mathematics, with 76 percent scoring in the upper half of the scale. Japanese students also were the most positive in their liking of mathematics, American youths the most negative [36]. The performance of 13-year-olds and of final-year secondary school students in Japan and the United States are compared in Fig. 21-5. Since the comparison is of scores earned 19 years ago, these students are now in their prime years in the labor force [37]. For countries competing in the international market with products and services that depend on advances in modern technology, these results are significant.

Technical and junior colleges

There are two types of mid-level institutions in Japan—technical colleges and junior colleges—that educate technicians [38]. Although both began shortly after World War II, they were started for different reasons and have come to serve different functions.

Shortly after the economic recovery began, Japanese industry recognized the need for mid-level technicians highly trained in practical applications of science and technology. In 1962, the Ministry of Education created 19 five-year technical colleges that combined the three years of upper-secondary schooling with two years of college. Among their purposes was to provide basic knowledge and techniques of design, production and construction, as well as the ability to plan and execute work in areas of management, testing, research and surveying [39]. Although the number of technical colleges rapidly increased to 54 by 1965 (to provide uniform distribution throughout Japan), their growth since then has been modest.

In 1982, 57 technical colleges enrolled 44,013 students and graduated 7,650. Although there are 20 recognized major disciplines at technical colleges, two-thirds of the students major in

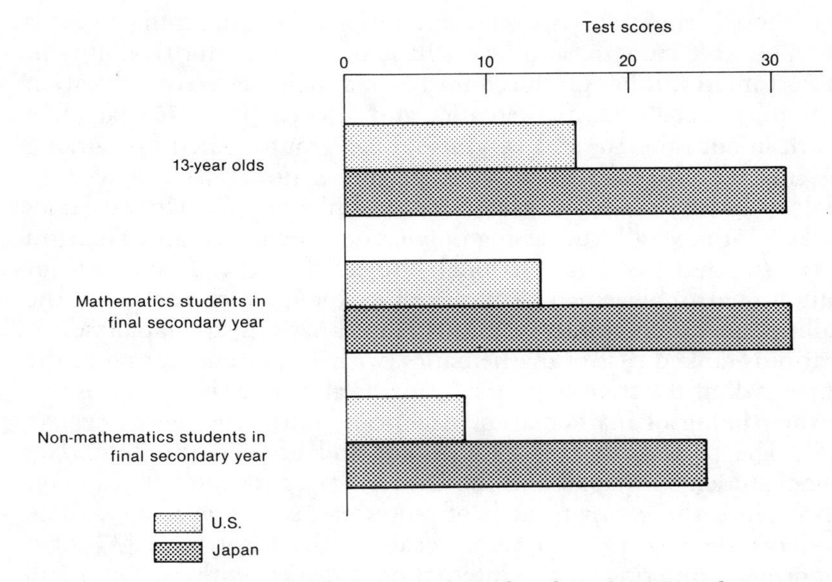

Fig. 21-5: *Mean mathematics test scores of students in U.S. and Japan.*

mechanical, electrical or civil engineering, with the remainder divided among aircraft, graphical and production engineering, industrial design, information engineering, and others.

In some ways, technical college courses resemble those taught at engineering colleges. In mechanical engineering, for example, there are courses on heat transfer, fluid dynamics, fluid machinery, automatic control, instrumentation and measurement, and dynamics of machines, as well as the properties of plastics and other non-metallic materials. The difference, however, is that the courses stress practical applications, rather than theory. Faculty members are recruited from among technicians in industry, as well as among persons with teaching experience in college and upper-secondary schools. Although technical college programs provide schooling only through the second year of college, they are similar in philosophy and operation to the four-year bachelor of engineering technology program in the United States. Technical colleges have been developed as a terminal level of schooling, but about 8 percent of the graduates continue their education at four-year colleges.

In contrast to technical colleges, which were established in response to industry demands, junior colleges were begun during the Occupation as part of the democratization process initiated by the Allies. Junior colleges, which are patterned after the two-year American model, are attended after completion of upper-secondary school. The Japanese educational establishment has been slow to accept the junior colleges, and they were not recognized under education law as permanent institutions until 1964.

In 1982, junior colleges graduated 6,222 people with engineering training, almost as many as the technical colleges. Although curricula are similar, junior college graduates are not highly esteemed in society. Technical college graduates receive a degree, while junior college graduates are awarded a certificate. Since many large companies will not hire persons with junior college training, they tend to become technicians in medium- and small-size companies that serve as subcontractors to large companies [40]. The low prestige of junior colleges may reflect the status of women in Japanese society. Of the 167,000 junior college graduates in 1982, over 92 percent were women, almost 30 percent of whom majored in home economics. Junior college education in Japan is considered a suitable preparation for marriage and homemaking, but not for industrial employment [41]. In industry women have had few opportunities for permanent employment or advancement, a situation that may gradually be starting to change.

The structure of higher education

Today, a dual system of higher education exists in Japan [42]. One group of institutions consists of low-tuition national universities and a very few high-quality private universities. These educate about 20 percent of the college population and usually are characterized by strong research activity and graduate courses. The second group is made up of a much larger number of high-tuition private institutions and local universities that educate the rest of the college population. Their main function is undergraduate instruction.

Differences between national and private universities are striking. As shown in Table 21-5, national universities surpass private universities by almost every quantitative indicator [43]. Very few of the best of the private institutions can match the resources of even the poorest of the national universities. The exceptions are Waseda, Doshisha, Keio and perhaps a few of the older private universities that enjoy a higher status and provide a better education than many of the newly created national universities [44].

Although 33 percent of Japanese upper-secondary school graduates pursue further education, compared to 42 percent of U.S. high school graduates, competition for admission—especially to the more prestigious universities—is keen. Tokyo Institute of Technology, for example, bases its admissions decisions not only on the applicant's

TABLE 21-5
COMPARISON OF NATIONAL AND PRIVATE UNIVERSITIES IN JAPAN, 1982

	National	*Private*
Number of institutions	95	326
Students	425,141	1,339,877
Full-time teachers	49,850	51,622
Part-time teachers	21,575	45,382
Ratio of students to full-time teachers	8.5	26.0
Floor space of school buildings (in thousands of square meters)	14,025	15,694
Size of school sites (in thousands of square meters)	1,334,932	125,193
1980 expenditures ($ in millions)*	4,262	5,331
Expenditures per student (actual $)*	10,025	3,979
1980 R&D expenditures in engineering ($ in millions)*	769.6	514.5
Baccalaureate graduates undertaking further education	15.2%	2.3%
Tuition	(Low)	(Moderate)

* Japanese figures converted at the 1982 exchange rate of 249.1 ¥ (yen) = $1.

performance on the national Joint Achievement Test, but on its own 6.5-hour series of written examinations in mathematics, physics, chemistry and a foreign language. Nationwide, in 1978, only 17.9 percent of the applicants to engineering schools were admitted. Of those, 72 percent had graduated from upper-secondary school that same year, while the rest were *ronin* (students who had failed to gain admission to a university in previous attempts). Of the *ronin*, 22 percent had graduated the previous year, and 6 percent two or more years earlier [45].

Engineering curricula

Colleges and universities in Japan continue to play an important role in economic development, having graduated more engineers in the last 25 years than have U.S. schools. Today, 20 percent of Japanese baccalaureate-level students graduate with engineering degrees, and another 3 percent graduate in natural sciences. In comparison, 4 percent of U.S. undergraduate degrees are granted in engineering, and another 7.5 percent in natural sciences. Currently 29 percent of Japanese engineering students are educated at national

TABLE 21-6
ENGINEERING DEGREES BY FIELD IN THE U.S. AND JAPAN, 1982

	Bachelor's		Master's		Doctoral	
	U.S.	*Japan*	*U.S.*	*Japan*	*U.S.*	*Japan*
Mechanical	21.2%	21.4%	13.9%	16.9%	11.8%	12.9%
Electrical & Electronics	24.0	26.4	23.1	22.7	19.0	21.2
Civil and Architecture*	16.1	24.4	16.7	15.0	12.7	13.0
Chemical	10.5	11.5	6.9	19.3	11.0	22.9
Computer	4.0	—	7.4	—	4.5	—
Other	24.2	16.3	32.0	26.1	41.0	30.0
Total	100.0%	100.0%	100.0%	100.0%	100.0%	100.0%
No. of degrees granted to:						
U.S. or Japanese citizens	61,580		13,258		1,720	
Foreign nationals	5,410		5,285		1,167	
Total	66,990	73,593**	18,543	7,363**	2,887	621*

* Architecture is not included in U.S. figures.

** Almost all graduates are Japanese citizens; in 1981, of the 414,165 students studying engineering or science in Japan, only 1,629 were from abroad.

universities, 69 percent at private universities, and less than 2 percent at local universities.

Students in Japan are distributed among engineering disciplines about as in the United States (Table 21-6). The Japanese, however, graduate a substantial number of people with concentrations in engineering management or administration backgrounds. Unlike U.S. schools, the Japanese include architecture as part of civil engineering.

In curriculum matters, the requirements for a bachelor's degree in engineering in Japan resemble those of U.S. schools. This similarity is not surprising, since the major educational reform occurred after World War II. The Allied occupational forces brought in groups of American advisers, including one engineering education group established by the American Society of Engineering Education, under the chairmanship of Harold Hazen [46]. As the Japanese attempted to follow the lead of the United States in technological development, many Japanese students have studied in the United States since the 1950s and brought back to Japan the best of what they found.

Degree requirements for a few Japanese and U.S. universities are given in Table 21-7. Although requirements vary from institution to

TABLE 21-7
COMPARISON OF REQUIREMENTS FOR B.S. DEGREE IN ENGINEERING IN SOME U.S. AND JAPANESE UNIVERSITIES

	Univ. of Tokyo	*Tokyo Inst. of Tech.*	*Waseda Univ.*	*Mass. Inst. of Tech.*	*Univ. of Illinois (Urbana)*
Institutional Characteristics					
Type of institution	Nat'l	Nat'l	Private	Private	State
Undergraduate students	1,815	3,287	40,800	4,700	25,400
Graduate students	1,340	1,659	3,200	4,700	7,300
Requirements					
Humanities & social sciences	24	24	48	24*	22
Natural sciences	12	18		20	36
Foreign language	16	12	14		
Departmental courses	84	70	80	60	45
Electives				16	25
Physical education	4	4	4		
Total credits	140	130	146	120	128
Annual tuition**	$816	$653	$2,449	$7,400	$634

* Three units are considered equivalent to one credit for purposes of comparison.
** Japanese figures converted at the 1981 exchange rate of 220.5 ¥ (yen) = $1.

institution, the three Japanese schools selected are probably as representative of universities in Japan as the U.S. institutions selected are of universities in this country. The Japanese requirements are similar to those in U.S. schools, with one major addition—the Japanese require 12 or so more credits, which are devoted to the study of one or two foreign languages. This training, following six years of foreign language study in secondary school, gives Japanese professionals a major linguistic advantage over their American counterparts, and allows them to read the literature of at least the English-speaking countries. This facility is capitalized on at the universities. Of the 276,000 volumes in the University of Tokyo's engineering library, for example, only 111,000 are in Japanese or Chinese. Waseda University has a library of over one million volumes, of which slightly more than 30 percent are in Western languages.

This language ability fits well with the nation's approach to development. Since the Japanese historically have pursued the course of primarily adapting western technologies, rather than relying on original research and innovation, they must know what other countries are doing. Japanese ministries, large industries and trading companies expend a huge effort in information gathering and analysis [47]. Vogel has described the scope and importance to the Japanese of these activities:

> If any single factor explains Japanese success, it is the group-directed quest for knowledge. In virtually every important organization and community where people share a common interest, from the national government to individual private firms, from cities to villages, devoted leaders worry about the future of their organizations, and to these leaders, nothing is more important than the information and knowledge that the organizations might one day need... It is not always clear why knowledge is needed, but groups store up available information nonetheless on the chance that some day it might be useful [48].

Graduate schools were not well developed before World War II. They existed, nominally, but there was no prescribed program and no fixed period of residency for candidates. Students pursued their own research interests under a senior professor, while they awaited appointment to the faculty [49]. After the war, under American influence, Japanese universities began offering master's degrees in the mid-1950s and doctorates in the early 1960s. The growth of graduate programs has been slow, however, as industry has not shown a keen interest in hiring people with doctorates.

Although Japan graduates substantially more engineers at the bachelor's level than does the United States, the situation, as shown in Figs. 21-6 and 21-7, is quite the reverse at the graduate level. In 1963, Japan graduated 693 engineers with master's degrees. By 1982 that figure had risen to 7,363. These numbers are far lower than in the United States, where over the same period the number of master's degrees awarded increased from 9,666 to 18,289 (of which 13,258 went to U.S. citizens or permanent residents). At the doctoral level, the difference is even more pronounced. From 1963 to 1982, the number of engineering Ph.D.'s granted in Japan increased from 83 to 621, while in the United States the number increased from 1,385 to 2,887 (of which 1,720 went to U.S. citizens or permanent residents). Even with the small number of graduates, employment opportunities for Ph.D.'s in Japan are limited. In 1982, the Ministry of Education reported that of recent graduates in engineering, one of 46 baccalaureate-level and one of 58 master's-level graduates were unemployed, but one of every seven Ph.D. holders had not found employment [50].

University departments (called faculties), which are collections of chairs that have been grouped for administrative, teaching and

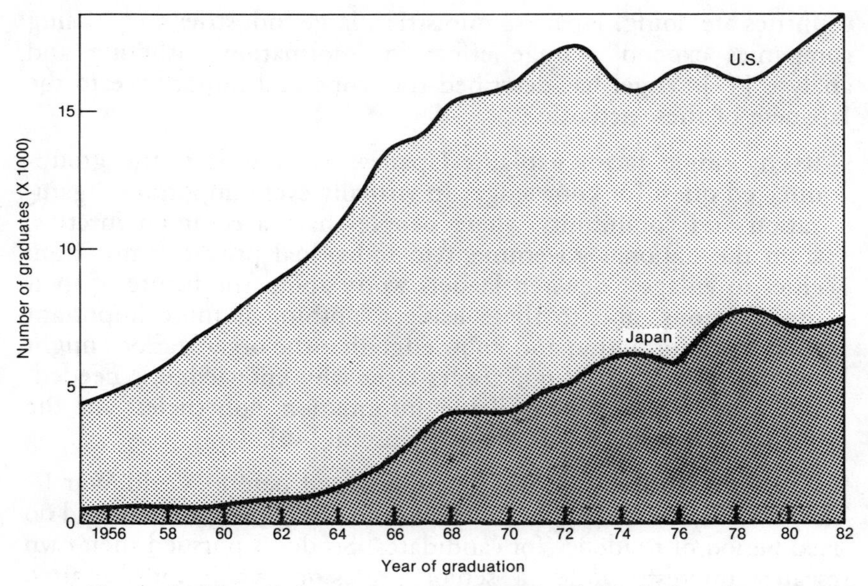

Fig. 21-6: *Master's degrees in engineering awarded in U.S. and Japan.*

research purposes, are relatively small but numerous in Japan. A university frequently will have several faculties with similar names that teach the same or similar subject matter. Each chair is an administrative unit consisting of one professor, two or three assistant professors, and several lecturers, assistants and technicians, all selected by the professor but approved by the faculty conference (i.e., departmental assembly). A chaired professor holds absolute authority over those under his control. They must totally support his interests; his betterment means the betterment of the group.* This structure is reinforced by the strong group orientation of Japanese society. Young faculty members defer to their elders in virtually all matters. Chair-holding professors closely and restrictively control the granting of the Ph.D. As competition among chairs for funds, space, new assistants and the most able students is intense, and since the success of the group depends on how well it works together, isolation and inbreeding result.

The lifetime employment system means there effectively is immediate tenure upon being hired as an instructor. There is also virtually no movement of faculty members to universities considered

* This structure reflects German influence in an early period of scientific development when a single professor had the knowledge of an entire field, such as organic chemistry.

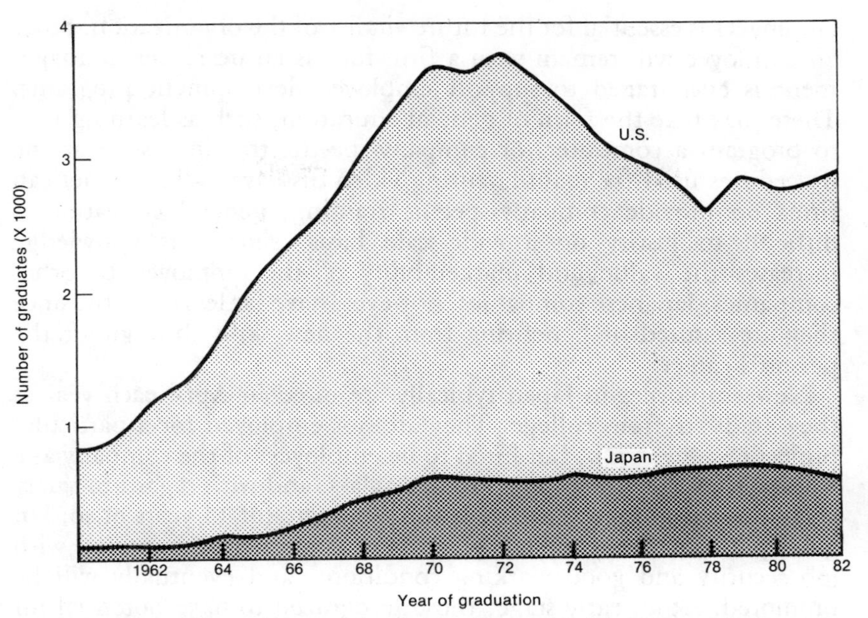

Fig. 21-7: *Doctoral degrees in engineering awarded in U.S. and Japan.*

less prestigious than their current one. As one can rarely expect to be employed by a university higher in prestige than one's alma mater [51], there is great inbreeding at the universities of highest esteem. In the late 1960s, 95 percent of the professors at Tokyo University and 89 percent at Kyoto were alumni of those institutions. Even at newer national universities, 40 percent of the assistant professors were alumni of the same institution in which they taught. The less prestigious institutions must rely on younger faculty members, or on older ones who have retired or have not been accepted into prestigious institutions. Graduates of Tokyo and Kyoto Universities thus have come to occupy 38 percent of all university positions in Japan [52]. Further, laws prohibit any non-Japanese person from holding a regular faculty position in a national university. These factors are serious for both the teaching and research functions of the university, for productive scholarly activity requires the free flow of ideas and information.

Training in industry

Under the lifetime employment system of large companies and government agencies, upgrading the qualifications and skills of employees is essential for the future vitality of the organization. Since an employee will remain with a firm for his entire career, management is encouraged to support employee development programs. These may take the form of general education, such as learning how to program a computer, or company-specific training, such as the procedures used for manufacturing video displays. While American firms do provide company-specific training, general education is difficult to justify on a wide-scale basis, since this knowledge increases the value and marketability of the employee to other companies. Japanese companies, however, have far less concern, since they are assured of benefiting from the new skills throughout the person's career.

New employees in Japan typically are hired in April each year as they graduate from college. They are not employed for a particular project or a given job, but hired to be employees of the company as a whole. They are viewed as a group or class, and with regard to salary and titles they progress at about the same rate for 15 years or so. For good performance, they are given new assignments, provided with job security and good working conditions, and eventually will be promoted. At an early stage, those recognized to have potential for high-level management are given assignments, but not necessarily the title or salary, to develop their potential. Eventually, they rise to high-level positions. The lengthy period of equality not only binds the group together, but allows ample time for employees to develop and demonstrate their abilities. By the time they reach their late thirties or early forties, the class begins to spread, as some of the members rise into middle and senior management. When they are about 55, most members of the group will retire and only those who have been identified for top executive levels of management remain.

Japanese companies have not been inclined to hire people with advanced degrees, particularly the doctorate. They prefer to hire baccalaureate graduates and train them to meet company goals. This practice has several advantages for the company. First, it is believed that young graduates are more malleable than older people with an advanced degree. The orientation and early training programs are expected to have more effect on younger graduates in helping them to identify with the company. Second, the training can be specialized to meet the company's specific needs, particularly in technical areas. Finally, the person with training provided by the company lacks the advanced credentials that might make him more employable on an open market and thus less committed to the company.

No universal pattern or policy for employee development or continuing education is followed by all companies. Each Japanese company is free to develop its own priorities and policies, consistent with its own objectives and knowledge of good business practices. As one would expect, large companies operate their own education and training activities, some quite extensive. Nippon Electric, for example, provides continuing education for its professional employees at its Institute for Technical Education. Hitachi has established the Hitachi Institute of Technology, where more than 5,000 company engineers were trained in the last ten years. Medium and small companies, however, must rely on business organizations or associations for educational programs.

Matsushita is one company that invests heavily in employee development. Virtually every one of the over 120,000 employees has had substantial company-provided training. The firm makes no particular effort to recruit from the elite universities, in contrast to most Japanese companies, as it believes in hiring young people who can be easily trained and starting them at the bottom. Every professional, whether engineer, accountant or salesman, starts by spending six months in training status, which includes selling or working directly in a retail outlet, in a factory performing routine tasks on an assembly line, and attending lectures by senior executives on "the Matsushita way." After the initial training, employees are assigned to divisions of the company, where they continue to take courses from its Research and Training Institute. Mentoring and job rotation to gain a broader overview of the company figure in the employee development program. When professional or managerial employees are to be promoted, they will study at the Institute for almost a year to upgrade technical and managerial skills, and for reinforcement in the company spirit [53]. Every employee is taught the corporate philosophy, organization, procedures and management system. These shared understandings and common belief that their most important activity is to meet customer needs with marketable products at minimum cost has been extremely important to Matsushita as it has grown and diversified its product lines [54].

Some implications

Japan has developed an educational system that has served its industrial needs well for the last 30 years. Guidelines from the centralized Ministry of Education, Science and Culture, frequently developed in response to the nation's economic plans, and a strong reliance on the nationally administered Joint Achievement Test as the primary factor in determining college admissions, have fostered uniform and high standards among the elementary and secondary schools of the country. Graduates of the secondary schools are well-grounded in science, mathematics and other subjects, and provide Japanese industry with a well-educated labor force receptive to further training at the workplace. Technical colleges and junior colleges graduate the technicians for large- and medium-size companies, while four-year colleges and universities educate engineers and other professionals. Industry provides continuing education and training throughout the working life of an individual both to increase his general intellectual growth and to make him more knowledgeable about the needs and procedures specific to the company. Education, in its entirety, has provided the intellectual base necessary for the rapid and large economic growth of the country.

Conditions in Japan, however, are changing. Japanese industry has become the leader in several fields of high technology, primarily by importing basic knowledge and ideas discovered in other countries and using them to develop and manufacture new or improved products. This approach may not continue as a fully successful strategy for the future, as the Japanese will be required to develop their own technologies. Continued economic growth may demand significant changes in Japan's present strategies for technological development and, in turn, to its educational system.

Part III—The future

Japan has been successful in building an economy to a level and at a rate unmatched in modern economic history. Since the end of World War II, Japan's gross national product has grown almost 60-fold, to a per capita level second only to the United States. Today, Japan accounts for about 10 percent of the world's economic activity, although occupying only 0.3 percent of the world's surface and supporting about 2.6 percent of the world's population. Can Japan in 1984 continue the growth of the past 30 years? If the nation is to progress, what implications does this have for education and training in Japan?

The remainder of this decade is significant for the Japanese. International responses to the nation's level of exports, the maturing and stabilization of its economy, and the rapid aging of its population and work force are trends that may require significant changes within Japan, if the nation is to maintain its economic vitality. These changes will probably occur in its business practices and social structure, its views on research and its educational system.

Japan's success in exporting goods is directly affecting other nations' industries. As a result, many countries are beginning to place limitations or restrictions on the import of Japanese goods.

France recently has required all Japanese electronic products to enter the country through Poitiers, a small port city, causing long delays in clearing customs and reaching French markets. The United States has negotiated voluntary restrictions on the number of Japanese cars that can be exported to the United States in any year. Several countries have imposed high import duties on Japanese goods. Japan may have to moderate its exports if further trade barriers are not to be imposed. As Japan relies heavily on exports, this will directly affect its economy.

Japan's economic growth is beginning to stabilize. For the 15-year period from 1955 to 1970, the average annual increase of the nation's real GNP was 10 percent. For the decade of the 1970s, however, the average annual growth rate was slightly less than 5 percent [55], and Japanese economists predict that the nation's economy will continue to grow at about this same rate for the remainder of the decade. The predicted world economic growth rate for the same period is slightly under 4 percent [56]. A more stable economy implies fewer job openings and less need for new, young workers in Japanese industries.

The population is aging at a rate unprecedented among modern nations. Japanese longevity has increased from an average life expectancy of 50 years in the mid-1930s to 70 years today. To exacerbate matters, since the "baby boom" after World War II, fewer children are being born to bear the social and economic costs of an aging population. In 1950, only 5 percent of the population was 65 years of age or older. Today, about 9 percent of the population is over 65 years of age, and that group will increase to 14 percent by 2000 and about 20 percent by 2020. In the United States, by contrast, persons over 65 years of age constituted about 8 percent of the population in 1950, and about 12 percent today. This group, however, will remain at this percentage until after the year 2000 and then rise to about 14 percent by 2020 [57]. In Japan, 15 working people now support each senior citizen; by 2015 it will be just three [58].

The work force also is aging rapidly. Under the traditional Japanese wage structure, in which salary is correlated with age, this implies higher average wages and greater labor costs, raising the prices of products and making them less cost competitive in international markets. The traditional system was beneficial to industry in the 1950s, when one-quarter of all workers were under the age of 25. The rapid expansion of industry and the influx of new young workers kept the percentage of employees under the age of 25 to 21.6 percent of the labor force as late as 1970. Slower growth, however, lowered the proportion of young workers to 13 percent of the labor force by 1978, and the average age of employees continues to rise [59]. Matsushita Electric Industrial Co., for example, estimates that if there is no change in its pay system, the aging work force will cause its payroll to rise 30 percent in the next ten years [60].

Future directions

How should Japan change to account for these trends? It is a country with few natural resources, that currently imports a majority of its energy, and that relies on other nations for an increasing share of its food. Since Japan cannot depend for its future economic growth on agriculture or the manufacture of goods whose production is energy-intensive, labor-intensive or consumes large quantities of natural resources, it must rely on the development of advanced technologies. One official of the Ministry of International Trade and Industry (MITI) has said that in the future, "*All* industries have to become smarter both in the way they make things and in the amount of knowledge the products themselves contain [61]."

The strength of Japanese industry has lain in adopting fundamental knowledge from other countries, improving on it, and designing, manufacturing and marketing the resulting products. The success of this approach, particularly in the high-value-added, high technology industries, has rested on a well-educated workforce.

To obtain the basic knowledge it requires, Japan has relied on licensing agreements. In almost all of the past 30 years and in each year since 1975, Japan has increased over the previous year the amount of technology imported [62], with over 50 percent of the technology introduced into Japan coming from the United States. As the state of technology changed, the types of Japanese imports changed. In 1975, for example, 9.6 percent of the agreements were for computer hardware and software; by 1980, 22.2 percent were for the same items.

Japan's strategy of identifying and transforming existing fundamental knowledge into products for the marketplace has been eminently successful. In early 1970, sales of products derived directly from imported technology accounted for more than 30 percent of industrial sales in Japan, and since that time imported technology has contributed increasingly to the nation's production of goods for export [63]. Japan, however, can no longer rely primarily on the United States or Europe for basic knowledge. As Japan competes in fields such as semiconductors, computers, telecommunications and genetic engineering, where the state-of-the-art is changing rapidly and in which organizations that do the basic research and development have a competitive advantage, its strategies will have to change. For the first time, Japan is in the position of having to advance the state of knowledge, do advanced research and create its own technologies.

The Industrial Structure Council, a policy advisory body to the Ministry of International Trade and Industry, foresees that Japan must develop a creative, more knowledge-intensive industrial structure. It must be based on the capability for original technological developments and produce higher-value-added products by focusing on software and knowledge intensification [64]. A Nikko Research Institute report on potential industrial growth areas for the 1980s stated the need as follows: "Japan has to develop its own technologies by means of the superior brains of its people—its only resource—in order to ensure its economic survival [65]."

In short, the nation must rely on its "brainpower." That is the intellectual challenge that faces Japan, and which Japan poses to other countries.

Need for creativity

Creativity is essential for technological leadership. The Japanese recognize this and are intent on fostering innovative thinking. Takuma Yamamoto, president of Fujitsu, summarized the views of many of his countrymen with his statement:

> The creativity of the Japanese people will be called into question from the latter half of the 1980s through the 1990s. The whole nation must work like one possessed to meet this great challenge [66].

In the past, however, there have been few examples of Japanese creativity in original technical development. A 1976 report of the National Science Foundation noted that Japan developed only 34 of 500 significant technological innovations introduced from 1953 to 1973, only 7 percent of the total [67], while the United States was credited with 63 percent of those innovations. The areas in which Japan has done best, such as semiconductors, VLSI, fiber optics and carbon fibers, do not reflect striking originality. In contrast to the United States, where there is strong emphasis on originality, Japan has stressed pragmatic thinking in its R&D, seeking product quality and reliability.

This approach is consistent with Japan's social orientation. Group identity and harmony are extremely important in Japanese society, influencing not only business practices and organization, but family relations, religious practices, social organizations and interpersonal relations [68]. Computer specialist Yasuo Kato has stated that the Japanese "are not so creative because the creative mind is peculiar, and we Japanese don't like anything peculiar. We believe that everyone should be the same [69]."

Social uniformity, however, need not preclude change. Japan has shown itself on several occasions to be highly adaptive and has adopted practices that have conflicted with the existing social system. When the Japanese were awakened to the advances of the West by Commodore Perry's visit, they rapidly changed from a feudal, agricultural society closed to outsiders to one with a manufacturing and industrial base driven by expansionist ambitions. After the nation's defeat in World War II, the Japanese readily adopted American ideas and institutions, modifying them to fit the nation's needs, and a period of rapid economic growth began. There is no inherent reason why the Japanese cannot again modify their social system, if the need is strong enough.

Changes in education

Perhaps nothing is more fundamental to Japan's technological development and economic growth in this century than education. The Japanese have created an extremely fine educational system that has produced a highly literate population, very homogeneous in its basic knowledge, with a good understanding of mathematics, science and other subjects central to modern technology. Yet there are significant problems in the educational system, particularly at the university level.

Fostering creativity, for example, is intertwined with questions of education. If Japan is to develop personnel capable of achieving technological breakthroughs, it must stress individuality, research, originality and risk taking. These are not the characteristics currently promoted in education. The significance of the single university entrance examination creates great pressure on the student to learn the types of information included on the examination. This has the effect, as Reischauer describes, of "distorting the content of his education [70]." Much of senior high school is devoted to preparing students for the university entrance examination, rather than to learning in the broader sense. Former Minister of Education Michio Nagai states, "It is not an exaggeration to say that education designed to develop men who love learning and think for themselves has already been abandoned [71]."

Although a certain amount of local autonomy is allowed at the elementary and secondary levels, the published curriculum standards and control of the certification of textbooks by the Ministry of Education promote a condition of uniformity in which children in the same grade, all over Japan, learn the same *katakana* characters and the same way to solve equations at almost the same time. The Japanese classroom, even at the university level, is characterized by rote learning, copying of lecture notes and an almost absolute reliance on testing and test results. In Japanese schools, says a Tokyo University professor, Takemitsu Hemmi, "students don't have to be able to discuss. They just say, 'Yes, I understand' [72]." The result is students who learn great amounts of information and typically far outperform students from other countries when tested in traditional ways. This method of teaching is probably the reason for the very significant increase noted in recent measurements of the IQ of the Japanese [73].

In spite of the rapid assimilation of large amounts of information—or perhaps because of the system that encourages it—the Japanese have not been noted for creative and original thinking. Masanori Moritani, a senior researcher at Nomura Research Institute, finds that:

> What Japan hungers for today is not a uniform crop of gifted students but people with extraordinary talents, heterodox people, human resources with the potential for achieving great things ...we [the Japanese] are entering an age when we must strive to search out and pinpoint the extraordinary and unorthodox among us... there is a real danger that these individualists will be numbered among the dropouts in today's education system. The question of how Japanese industry can best discover and train men and women of this disposition will be one of the greatest issues facing it in the future, on which will hang the success or failure of growth and development for years to come [74].

Higher education has played less of a role in Japanese society than the entrance examinations would suggest. After the intense studying to pass these examinations, students are granted a respite from continued pressure. They are rarely failed or academically challenged to the same degree as in high school. Faculties are highly autonomous and maintain a great deal of power in most universities. They protect their budgets and academic domains tenaciously. Curricula are rigidly prescribed, new fields of study are difficult to start, and cross-disciplinary studies are almost impossible to have approved. There is little opportunity for innovation in universities.

Although recruitment of university graduates for government and business positions usually is by examination, there is a strong correlation between one's performance on these examinations and the university one attends. This probably is due more to the native ability of the students and to the fact that the largest firms limit their recruiting to the most prestigious universities, than to the quality of students' university studies [75]. Those who score highest on employment examinations are most likely to be the ones who, four years earlier, scored highest on the university entrance examinations. The fact that the needs of Japanese industry are met as well as they are, in spite of the problems of higher education, is due in large measure to the high quality of pre-university education and to the pre-service training programs conducted by industry and government.

Recognizing that education will continue to be an essential element in Japan's strategy for development, MITI has proposed that for the future well-being of the nation, Japan must establish *an age of vitalized human potential* [76]. Although a generally accepted goal, it is not clear what this means or how it will be achieved. There is already a trend in Japan toward higher levels of education for both men and women. Since 1970, persons over 20 years old who have completed higher education have increased from 11.5 to about 16 percent of the population. By 2000, one in four Japanese is expected to have completed post-secondary education [77].

Although there has been a great increase in the number of students receiving baccalaureate degrees, few people complete graduate studies. Advanced graduate education currently is undertaken primarily by those preparing for academic careers. As noted in Figs. 21-6 and 21-7, Japan is graduating about half the number of master's and one-third the number of doctoral candidates in engineering as is the United States. In spite of the difference in numbers, the United States cannot meet the demands of its academic institutions for persons with doctorates, while there is very little demand in Japan for those educated to this level. The situation, as shown in Fig. 21-8, is similar in the sciences [78]. The total number of people receiving master's and doctoral degrees in physics and chemistry in Japan is only about one-fourth that of the United States. In its desire to base its economy on knowledge industries and achieve world leadership in high technology, Japan will require significantly more people with graduate education, particularly Ph.D.'s in engineering and science.

Research activities

To achieve its goal of becoming a knowledge-intensive nation, Japan will have to foster environments that can lead to sustained

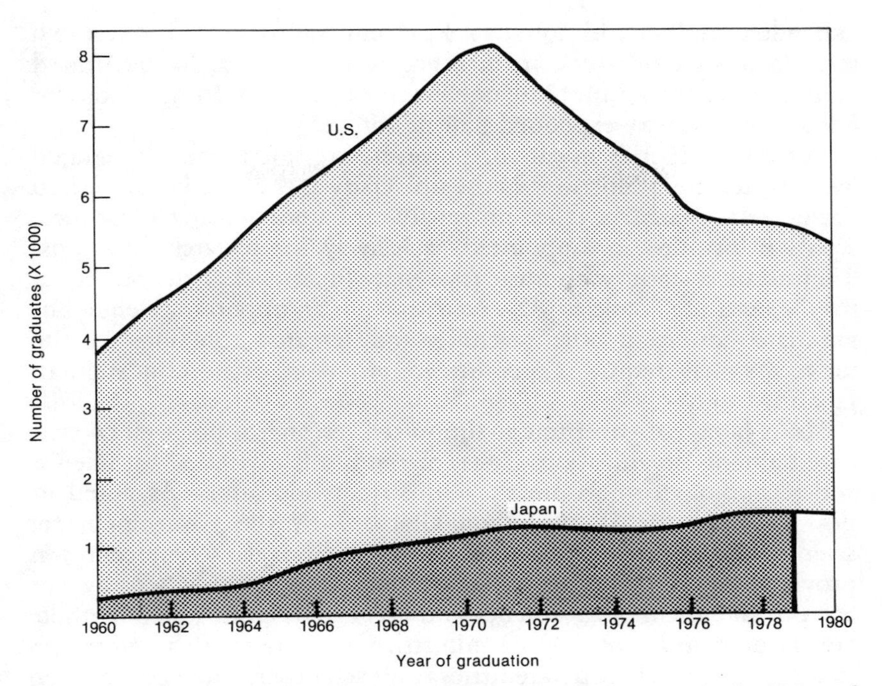

Fig. 21-8: *Master's and doctoral degrees in chemistry and physics in the U.S. and Japan. Engineering degrees follow the same pattern.*

success in extending the state of knowledge in many scientific and technological fields. The striking feature of Japanese research, however, is its direct relevance to the needs of industry and commerce. This is true of research supported by the government as well as business. Industry's concept of research and development is exemplified by Matsushita, where a primary purpose of its 23 production research laboratories is "to analyze competing products and figure out how to do better [79]."

The degree to which this utilitarian view channels Japan's creative efforts into activities with commercial possibilities, rather than extending the state of knowledge, may be gauged by an analysis of citations in the scientific literature. The Institute for Scientific Information included only 19 Japanese among the 1,000 scientists credited with the most frequent citations. Further, approximately half of these Japanese were affiliated with American universities [80].

University-based research activities in Japan are extensive. In 1980, $3.6 billion, which was 17.6 percent of the total R&D funds expended nationally, was spent at universities. Industry accounted for $13.9 billion, which was 67.1 percent of the R&D funds expended, and an additional 15.3 percent was expended at research institutes, some of which are attached to universities. In the United States, by contrast, in 1979 universities expended $5.4 billion for R&D, and industry expended $38 billion [81].

Research activities in Japan, however, are not as well integrated with the teaching functions of the university as they are in the United States. At many of the university-affiliated research institutes separate faculties are appointed exclusively for research functions. They attain a status within the university independent of, but equal to, the teaching faculties [82]. As research is an important element of graduate education, the teaching and research functions of the university may require closer integration if graduate education in Japan is to grow.

Much Japanese government-supported research is not well coordinated among organizations. Most of the funds expended for science and technology come from the Ministry of Education. As noted in Fig. 21-9, of the $6.53 billion allocated by the government for science and technology in the FY 1983 budget, 49 percent was provided by the Ministry of Education, and notably less by the Science and Technology Agency and MITI [83]. Unfortunately, there are strong rivalries among the ministries, and many of the ministries that support R&D cooperate little with each other. They do not even relate well to the Science and Technology Agency, which was established to plan and coordinate the government's efforts in these areas [84]. Space research, for example, is conducted at the University of Tokyo under support from the Ministry of Education. Space applications R&D is carried out by the National Space Development Agency, under the direction of the Science and Technology Agency. The two organizations have separate launch facilities and do not engage in any interchange of scientific personnel [85].

The pattern of research support is quite different between national and private universities in Japan. Almost all of the funds at national universities are provided by the government (Table 21-8), while 83 percent of the engineering research funds expended at private universities are provided by industry and other non-government sources [86]. Although it would appear that private universities in Japan have stronger links with industry, it must be recalled that the national government plays a strong part in industry planning and coordination. The national universities can serve a valuable function in a three-way partnership as the government attempts to develop future economic opportunities for the country.

Japan recognizes the need to stimulate its creativity and already is supporting basic research in areas such as VLSI, fiber optics, communications systems, voice recognition systems and other ad-

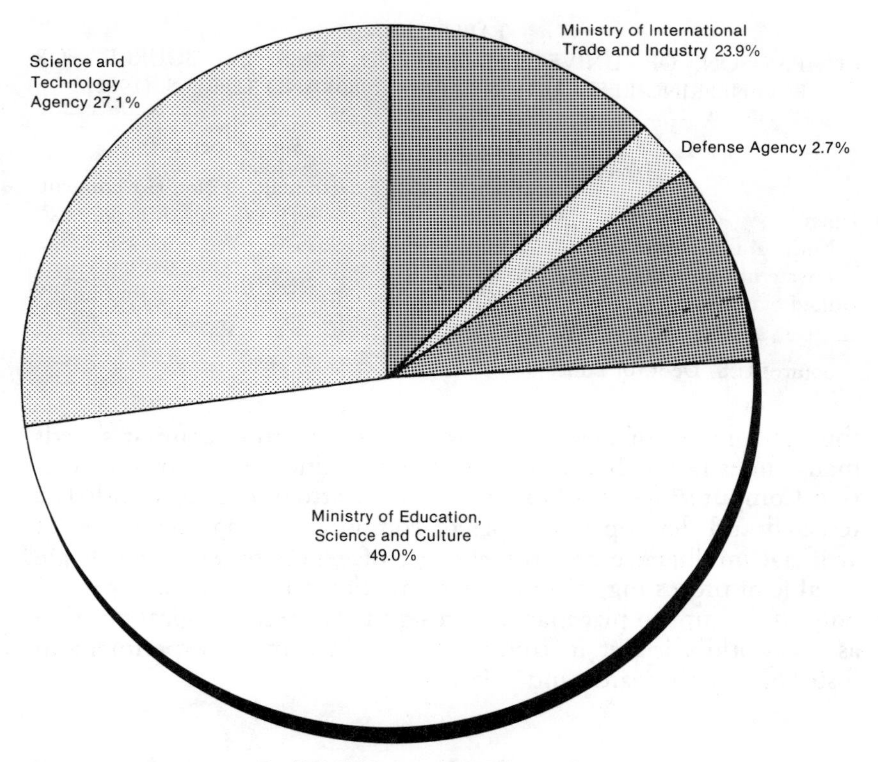

Fig. 21-9: *Japan's FY 1983 budget for science and technology by ministry and agency. "There are strong rivalries among the ministries, and many that support R&D cooperate little with each other."*

vanced electronic technologies, as well as in such urgently needed areas as energy conservation and pollution control. Further, the Industrial Structure Council of MITI has set numerous objectives for the 1980s, including both the further development of knowledge-intensive and innovative technologies, such as microcomputers, optical communications, VLSI and laser beam devices, and the creation of next-generation technologies in the life sciences, energy and data processing. Particular emphasis will be on creating new materials, developing large-scale systems for alternative energy sources and creating technologies for social systems, such as for personal and community activities [87].

To stimulate creativity on a more national scale, the government and industry are jointly supporting two major activities. The National Superspeed Computer Project is an attempt to develop computers

TABLE 21-8
COMPARISON OF UNIVERSITY EXPENDITURES BY SOURCE FOR ENGINEERING RESEARCH IN JAPAN (1980) AND THE U.S. (1982)

	Amount	*Source* (percent)	
	(in millions)	Government	Non-Government
Japan			
National universities	$766	98%	2%
Private universities	$512	17	83
United States	$1,000	81	19

Source: U.S. Dept. of Educ.

through the use of new materials and designs to operate at speeds many times faster than today's fastest machines. The Fifth-Generation Computer Systems Project is a major attempt to do world-class research and develop new types of computers, using techniques of artificial intelligence and the concept of expert systems, which are capable of processing information symbolically [88]. The latter is not only an attempt to place Japan in a significant technological position as the world's leader in computers, but is a major experiment in fostering basic research and creativity.

Changes in human resource development

Company-provided education and training play an important role in Japanese industry. Employees of Japanese firms are not hired to work at a particular job or within a particular unit. They are hired as employees of the company, who will work in a variety of jobs in various parts of the organization throughout their careers. With lifetime employment granted to many employees of large firms, vacancies at all but the lowest levels are filled by internal promotions. To be effective, this internalized labor market requires that workers not only obtain broad experience within the firm but receive appropriate training at various stages in their careers. This training is important both for motivating employees and for developing the needed human capital within the firm.

Significant increases in higher levels of education, the aging work force and the internationalization of Japanese business raise serious problems, however, for industrial education and training, and for the types of employment that will be available for Japan's already highly qualified work force. Greater numbers of college graduates are leading to a surplus and to underemployment; people are now having to accept jobs that previously were filled by those with less education. In 1978, almost 55 percent of the university graduates in industry in the 40 to 44 age group had advanced to the level of division or department manager; only 30 percent of university graduates in that age group are expected to attain management positions by 1988 [89]. The remaining 70 percent mostly likely will retire from non-management jobs [90].

If longevity no longer guarantees promotion, new forms of worker evaluation based on performance and new forms of education and training will have to be devised. Merit promotion systems have been used at companies such as Matsushita and Canon for many years, and are now being adopted by others. The notion of using examinations as the primary means of assessing an employee's ability is gaining currency. The examination system is seen as an objective method of detecting workers who may be lacking in a particular skill, which could then be provided through appropriate training and job assignments [91]. Although this approach to evaluation is a reflection of the examination system in schools, it is a marked change for industry in its procedures for personnel assessment.

Although the leading Japanese firms have developed elaborate training programs that emphasize total and comprehensive personnel development, most educational programs are provided for new employees who are recent graduates. Twice as many Japanese firms offer educational programs for the recent graduate than for either present employees or new employees who have prior experience with other firms [92]. This practice is in sharp contrast to the United States, where 90 percent of the expenditures for education in companies of over 500 employees is for present employees; only 10 percent is devoted to the education of new workers [93]. The majority of the jobs in Japanese industry were formerly designed for young workers, who constituted the majority [94]. Now tasks must be remodeled to fit the older worker. As this is not simply a question of finding a job, but of productively using the worker, it must be viewed within the larger context of career development and training.

As Japan's economy has grown, both its exports and its overseas investments have increased rapidly. Such investment is different from trade, for it involves conducting a business under local rules and regulations. In the larger companies, comprehensive training programs, which may extend over several years, may include overseas job rotation and instruction on the customs, culture, history, legal system, and political and social systems in the country, as well as language training [95].

Fujitsu has an interesting approach to its development program for businessmen to serve in international activities. Twenty percent of the management training they receive is devoted to cultural subjects, including music, art, drama, literature, history, religion and poetry, as well as international politics and economics. Since the early training and experience of the employee has been devoted to improving technical competence, the company now stresses subjects that will give him the personal skills and breadth of view needed to deal both formally and informally with people from other countries.

The future role of women

One of the most significant changes in Japanese society stemming from the emphasis on technology and education will be the assumption of new roles by women. Although women make up one-third of the labor force in Japan, they have occupied primarily clerical, assembly or temporary positions. Their expendability has been a significant factor in allowing large corporations to maintain the lifetime employment system for men.

The degree of education and the aspirations of women, however, are changing. The proportion of college-educated women is projected to increase from 6.4 percent of the female population in 1970 to 20 percent in 2000 [96]. Higher levels of education, coupled with a labor shortage in certain career specialties, are creating new opportunities for women. The Industrial Structure Council, for example, estimates that by 1985 Japan will require 795,000 software engineers. Today, there are fewer than 100,000 of these specialists in the country, and high-level male engineers are reluctant to remain in the field because in industry generalists have better promotion possibilities [97]. Industrial recruiters are beginning to hire software engineers from women's colleges.

Venture capital firms, although far fewer in Japan than in the United States, are also creating new options for women. Entrepreneurial firms have great difficulty recruiting highly qualified men, who prefer to enter large firms and rarely leave them once they are hired. Women, regardless of their qualifications and ambitions, have had almost no opportunities in large firms to rise to management positions—jobs reserved exclusively for men. Women are, however, being given management roles in new venture companies [98]. With significant numbers of women entering the professional work force, there will be pressure to have more women advance to management positions in small and large firms. This pressure will exacerbate a projected shortage of management positions for university graduates and will no doubt create new relationships between men and women. The effects of these relationships will extend well beyond the firm.

The changes beginning in many areas of Japanese society will accelerate as Japan continues to reach a more stabilized but advanced level of economic development. Prime Minister Yasuhiro Nakasone spoke to the extent of those changes when he said, "We [the Japanese] must formulate a society for which there is no precedent in any other country [99]." The future of Japan will depend on how well the nation can effect the changes it plans and deal with the desired and undesired consequences of those changes.

Postscript

There are contrasting aspects to relationships between the United States and Japan. On one side, the development and sale of manufactured products by Japanese firms have eroded traditional American markets, caused a decrease in revenue for our businesses and a loss of jobs for many Americans. In contrast, Japan is the cornerstone of American foreign policy in Asia, as well as a major trading partner. The mutual reliance is so strong that Saburo Okita, former Foreign Minister of Japan, was prompted to say, "Japan cannot survive without harmonious and cooperative relations with the U.S. [100]"

Bilateral trade between the two nations, which exceeded $59.6 billion in 1982, mushroomed from $1 billion in 1954 and $18 billion in 1973. Japan depends on the United States for its defense, as well as for a significant part of its food supply. Indeed, more acreage in the United States is devoted to producing food for Japan than within Japan itself. From America's perspective, Japan is the largest market, after Canada, for U.S. goods, and the best customer for agricultural products. Each country also is the other's largest overseas investment market. In 1979, Japanese firms invested about $2 billion in production facilities in the United States, creating an estimated 340,000 jobs in America. The United States in turn invested about $1.2 billion in Japan through a wide range of activities [101].

Japan will continue to be a formidable economic competitor, but the period when the nation could rely almost exclusively on improving ideas developed in other countries has come to an end. Japan must now stress individual creativity and initiative in basing its future industrial development on an intensified use of knowledge.

The beginnings of change are already evident. There is a growing national consensus that Japan should alter its policies concerning science and technology from emphasizing adaption to promoting creativity and inventiveness. The government is declaring that Japan will become a technologically-oriented nation, rather than a trade-oriented one. As it moves towards becoming a technology-based, knowledge-intensive nation, education will remain an essential element of Japan's development strategy. The Japanese will stress intellectual achievements over natural resources and physical labor as they strive to maintain economic vitality. The attitude of Japanese business towards education was expressed well by the president of Nippon Telegraph & Telephone Public Corporation (NTT) when he stated, "High intelligence is the only source of competitiveness [102]."

The United States can respond to the Japanese challenge by educating its citizens to live in, partake of and contribute to a high technology society. This task, however, will not be accomplished easily or quickly. A recent report of the National Science Board, in describing American education, notes:

> The nation that dramatically and boldly led the world into the age of technology is failing to provide its own children with the intellectual tools needed for the 21st century... Already the quality of our manufactured products, the viability of our trade, our leadership in research and development, and our standards of living are strongly challenged. Our children could be stragglers in a world of technology... We must not provide our children a 1960s education for a 21st century world [103].

As we study the situation in Japan, we should not feel compelled to mimic the Japanese. Rather, this review should be a way of reflecting on what we are doing and identifying the tacit reasons why certain developments have occurred in the United States. We should then be in a better position to challenge our assumptions, to identify and build on our strengths, and to change what we are doing in ways that fit out society's values and needs. As we view America's future, we should recognize that the United States and Japan are firmly linked as partners in a broad spectrum of regional and global interests. We can increase our understanding of the Japanese while strengthening our system to improve U.S. international competitiveness and maintain our position as the economic leader of the world.

References

[1] *Economic Report of the President,* U.S. GPO, Washington, D.C., Feb. 1982, pp. 279, 283.

[2] "International Comparisons of Manufacturing Productivity and Labor Cost Trends, Preliminary Measures for 1982," news release, USDL 83-248, U.S. Dept. of Labor, May 28, 1983, tables 2 and 8.

[3] "Japan's High-Tech Challenge," *Newsweek,* Aug. 9, 1982, p. 48.

[4] Allen, G. C., *A Short Economic History of Modern Japan*, Fourth Ed., New York: St. Martin's Press, 1981, pp. 2–3.
[5] Galbraith, J. K., *The New Industrial State*. Boston: Houghton Mifflin, 1971, p. 298.
[6] Kobayashi, T., *Society, Schools and Progress in Japan*. New York: Pergamon Press, 1976, p. 90.
[7] Shimahara, K., *Adaption and Education in Japan*. New York: Praeger, 1979, pp. 129–130; Kobayashi, *op. cit.*, ref. 6, p. 91.
[8] Kobayashi, *op. cit.*, ref. 6, pp. 91–92.
[9] Shimahara, *op. cit.*, ref. 7, p. 133.
[10] Kobayashi, *op. cit.*, ref. 6, p. 94.
[11] Although the scope of the Ministry's responsibilities have broadened since the 1960s, its major concerns are reflected in its current name as the Ministry of Education, Science and Culture. Although direct comparisons are imperfect, today the Ministry has responsibility for activities that in the United States would be under the Department of Education, the National Science Foundation, the National Endowment for the Humanities, the National Endowment for the Arts, and the Smithsonian Institution.
[12] Kobayashi, *op. cit.*, ref. 6, pp. 105–106.
[13] *The Effect of Government Targeting on World Semiconductor Competition, A Case History of Japanese Industrial Strategy and Its Costs for America*, Semiconductor Industry Assoc., Cupertino, CA, 1983, p. 39.
[14] *Ibid.*, p. E3.
[15] "International Comparisons of Manufacturing Productivity and Labor Cost Trends, Preliminary Measures for 1982," *op. cit.*, ref. 2, table 2.
[16] *U.S. Industrial Competitiveness, A Comparison of Steel, Electronics, and Automobiles, Summary*, Office of Technology Assessment, Washington, D.C., 1981, p. 12.
[17] Hosai Hyuga, chairman of Sumitomo Metal Industries, Ltd., Japan's sixth largest exporter of technology in 1980, states: "The high level and nature of the Japanese educational system makes it very easy to turn a high school graduate into an auto assembly line worker or a college graduate into an electronics engineer." Quoted in "Japanese Technology, The Cutting Edge," *Fortune*, Aug. 23, 1982, p. 24.
[18] Vogel, E., *Japan as Number One, Lessons for America*. Cambridge, MA: Harvard University Press, 1979, p. 45.
[19] *Open Doors, Report on International Educational Exchange*, Institute of International Education, New York, annual vols. from 1966 to 1981; *Profiles, The Foreign Student in the United States*, IIE, New York, 1981. Unpublished data for 1982–83 received from IIE. In obtaining data for 1979 and beyond, the Institute began two separate surveys. It continued the survey of institutions which yielded the data for 1970 and earlier, and began a survey of individuals to obtain the breakdown by discipline. The response rate for the latter was slightly over 50%, e.g., the institutional survey identified 12,260 Japanese students studying in the United States, while the survey of individuals identified only 6,842 students. The data by discipline for 1979–80 and 1982–83 has been scaled to make it more compatible with

previous years' data, and as such should be viewed for trends rather than precision.

[20] *Ibid.*

[21] Vogel, E., *Japan as Number One.* MA: Harvard University Press, 1979, p. 160.

[22] *Statistical Abstract of Education, Science and Culture,* 1983 edit., Ministry of Education, Science and Culture, Japan, 1983, pp. 84–86, 88.

[23] *The Condition of Education,* 1983 edit., Nat'l Center for Educ. Statistics, Washington, D.C., 1983, p. 58.

[24] Anderson, R. S., *Education in Japan: A Century of Modern Development,* HEW, 1975, p. 108.

[25] Fetters, W. B., *et al.*, "Schooling Experiences in Japan and the U.S.: A Cross-National Comparison of High School Students," unpublished paper presented at 1983 Annual Meeting, Amer. Educ. Res. Assoc., Montreal, Canada, Apr. 13, 1983, p. 6; Eckstein, M. A., K. S. Travers and S. M. Shafer, "A Comparative Review of Curricula: Mathematics and International Studies in the Secondary Schools of Five Countries," paper submitted to Nat'l Commission on Excellence in Educ., Apr. 28, 1982, p. 100.

[26] *Education in Japan, A Graphic Presentation,* Ministry of Education, Science and Culture, Japan, 1982, p. 59; Anderson, *op. cit.*, ref. 4, p. 112; Rosenshine, B. V., "How Time is Spent in Elementary Classrooms," *Time to Learn,* C. Denham and A. Lieberman, Eds., NIE, Washington, D.C., May 1980, pp. 110–111.

[27] Eckstein *et al.*, *op. cit.*, ref. 5, p. 31.

[28] Stigler, J. W., *et al.*, "Curriculum Achievement in Mathematics: A Study of Elementary School Children in Japan, Taiwan, and the United States," unpublished paper, Jan. 1983, p. 7.

[29] Eckstein *et al.*, *op. cit.*, ref. 5, p. 31.

[30] *Course of Study for Upper Secondary Schools in Japan,* Ministry of Education, Science and Culture, Japan, 1983, pp. 40–48.

[31] Eckstein *et al.*, *op. cit.*, ref. 5, p. 52.

[32] Hurns, C. J. and B. B. Burns, "An Analytic Comparison of Educational Systems: Overview of Purposes, Policies, Structures and Outcomes," paper presented to Nat'l Commission on Excellence in Educ., Feb. 1982, p. 57; Anderson, *op. cit.*, ref. 4, p. 167.

[33] Fetter *et al.*, *op. cit.*, ref. 5, tables 2 and 5.

[34] Cummings, W. K., *Education and Equality in Japan.* Princeton, NJ: Princeton U. Press, 1980, pp. 213–215.

[35] Vogel, *op. cit.*, ref. 1, pp. 159–160; Comber, L. C. and J. P. Keeves, *Science Education in Nineteen Countries, An Empirical Study.* New York: John Wiley, 1973, pp. 119–120, 159.

[36] *Science & Engineering Education for the 1980's & Beyond,* NSF and U.S. Dept. of Educ., Washington, D.C., Oct. 1980, p. 59; and Husen, T., Ed., *International Study of Achievement in Mathematics, A Comparison of Twelve Countries.* New York: John Wiley, 1967, vol. 2, pp. 22–25, 44–46.

[37] The second international assessment, involving 23 countries, is now being conducted. A discussion of preliminary results from Japan, where data were collected in 1981, a year before the other countries, is contained in L. P.

Grayson, "Leadership or Stagnation? A Role for Technology in Mathematics, Science and Engineering Education," *Engineering Education*, vol. 73, no. 5, Feb. 1983, pp. 356–366.
[38] Although the programs clearly are at the technician level, some authors have included the graduates in the total number of engineers, and have drawn the conclusion that about 15 percent of engineers in Japan are educated to a pre-university level. See, for example, J. Mintzes, "Scientific and Technical Personnel Trends and Competitiveness of U.S. Technologically Intensive Industries (A Comparison with Japan, West Germany, and France)," paper prepared for NSF, June 1, 1982; *Statistical Yearbook*, UNESCO, Paris, 1974, 1978, 1979.
[39] Anderson, *op. cit.*, ref. 4, p. 201; *Technical and Technological Education in Japan*, Japanese Nat'l Commission for UNESCO, Dec. 1972, p. 78.
[40] *Technical and Technological Education in Japan, op. cit.*, ref. 19, p. 62.
[41] Anderson, *op. cit.*, ref. 4, pp. 216–218.
[42] Shigeru, N., "The Role Played by Universities in Scientific and Technological Development in Japan," *Journal of World History*, vol. 9, no. 2, 1965.
[43] *Statistical Abstract of Science, Education and Culture, op. cit.*, ref. 2.
[44] *Higher Education and the Student Problem in Japan*, U. of Tokyo Press, Tokyo, 1972, p. 36.
[45] *Japan Statistical Yearbook*, Tokyo, 1980, table T.404, p. 593.
[46] Hazen, H. L., "The 1951 ASEE Engineering Education Mission to Japan," *Journal of Engineering Education*, June 1952, pp. 481–488. Hazen, H. L., "America Meets Japan in Engineering Education," *The Technology Review*, May 1952, pp. 351 ff. One result of the recommendations of the advisory group on engineering education was the establishment in 1952 of the Japanese Society for Engineering Education.
[47] Seiki Tozaki, president of C. Itoh & Co., Ltd., one of Japan's largest trading companies, says: "Today's trading firms are akin to giant think tanks active worldwide not only in the area of trade but also in development and investment. Through these three types of activities, they are able to collect the most outstanding software [i.e., knowledge] concerning product distribution, project investment and technology development. Expert use of this software thus enables *sogo shosha* [i.e., trading companies] to create and respond to all manner of business possibilities anywhere in the world. Searching out new technologies will receive the highest priority in their future activities." Quoted in "Japanese Technology, The Cutting Edge," *Fortune*, Aug. 23, 1982, p. 64.
[48] Vogel, *op. cit.*, ref. 1, p. 27.
[49] Anderson, *op. cit.*, ref. 4, p. 602.
[50] *Statistical Abstract of Science, Education and Culture, op. cit.*, ref. 2, pp. 96–97, 100–101.
[51] Cummings, W. K., I. Amano, and K. Kitamura, *Changes in the Japanese University, A Comparative Perspective*. New York: Praeger Publ., 1979, pp. 152–153.
[52] Anderson, *op. cit.*, ref. 4, p. 186.
[53] Olson, L., "Future Technical Manpower Requirements and Availability in

Information Technology R&D: United States Review and Comparison with Japan, France, and the United Kingdom,'' OTA, draft report, May 1983, pp. 64–65.
[54] Pascale, R. T. and A. G. Athos, *The Art of Japanese Management, Applications for American Executives.* New York: Warner Books, 1981, pp. 77–80.
[55] Long-Term Outlook Committee, Economic Council, Economic Planning Agency, *Japan in the Year 2000, Preparing Japan for an Age of Internationalization, the Aging Society and Maturity,* The Japan Times, Ltd., Tokyo, Jan. 1983, p. 14.
[56] ''The Vision of MITI Policies in the 1980s, Summary,'' report no. NR-226 (80-7), Ministry of International Trade and Industry, Tokyo, Mar. 17, 1980, p. 8.
[57] *Ibid.*, p. 7; Kikuchi, M., *Japanese Electronics, A Worm's-Eye View of Its Evolution.* Tokyo: Simul Press, Inc., 1984, p. 6; Shimada, H., ''The Japanese Employment System,'' The Japan Institute of Labour, Tokyo, 1980, pp. 29–30.
[58] ''The Social Impact of a Graying Population,'' *Business Week,* Apr. 20, 1981, p. 72.
[59] Shimada, *op. cit.*, ref. 3, p. 13.
[60] ''An Aging Work Force Strains Japan's Traditions,'' *Business Week,* Apr. 20, 1981, p. 81; ''A Changing Work Force Poses Challenges,'' *Business Week,* Dec. 14, 1981, pp. 116–117.
[61] ''Japan's High-Tech Challenge,'' *Newsweek,* Aug. 9, 1982, p. 48.
[62] Ozawa, T., *Japan's Technological Challenge to the West, 1950–1974: Motivation and Accomplishment.* Cambridge, MA: MIT Press, 1974, pp. 19, 23; *Indicators of Science and Technology, 1982,* Science and Technology Agency, Tokyo, 1983, p. 126.
[63] Caves, R. Z. and M. Vekusa, *Industrial Organizations in Japan,* Brookings Inst., Washington, D.C., 1977, ch. 7.
[64] ''The Vision of MITI Policies in the 1980s, Summary,'' *op. cit.*, ref. 2, p. 21.
[65] *Japanese Industries, New Technologies and Potential Growth Areas in the 1980's,* The Nikko Research Center, Ltd., Tokyo, Oct. 1980, p. 1.
[66] Moritani, M., *Japanese Technology: Getting the Best for the Least.* Tokyo: Simul Press, Inc., 1982, p. 184.
[67] Feinman, S. and W. Fuentevella, *Indicators of International Trends in Technological Innovations,* report for the National Science Foundation under contract no. NSF-C-889, Apr. 1976.
[68] Nakane, C., *Japanese Society,* University of California Press, Berkeley, 1972.
[69] ''Electronics Research: A Quest for Global Leadership,'' *Business Week,* Dec. 14, 1981, p. 89.
[70] Reischauer, E. O., *The Japanese.* Cambridge, MA: The Belknap Press, 1977, p. 175.
[71] ''Schooling for the Common Good,'' *Time* (special issue on Japan), Aug. 1, 1983, p. 21.
[72] *Ibid.*, p. 67.
[73] Lynn, R., ''IQ in Japan and the United States Shows a Growing Disparity,''

Nature, May 20, 1982, pp. 222–223; Anderson, A. M., "The Great Japanese IQ Increase," *Nature*, May 20, 1982, pp. 180–181. Lynn's results indicate that three quarters of the younger generation in Japan have higher IQ test scores than the average American or European. Further, 10 percent of these younger Japanese have IQ's in excess of 130, the level frequently found among professionals, compared to only 2 percent in the United States.

[74] Moritani, *op. cit.*, ref. 12, p. 179.

[75] Reischauer, *op. cit.*, ref. 16, p. 174.

[76] "The Vision of MITI Policies in the 1980s, Summary," *op. cit.*, ref. 2, p. 27.

[77] Long-Term Outlook Committee, *op. cit.*, ref. 1, p. 7.

[78] *Digest of Education Statistics;* U.S. Dept. of Education, Washington, D.C., annual vols. 1965–1982; *Earned Degrees Conferred by Higher Educational Institutions,* HEW, Washington, D.C., annual vols. 1960–1964.

[79] Pascale, R. T. and A. G. Athos, *The Art of Japanese Management, Applications for American Managers.* New York: Warner Books, 1981, p. 42.

[80] "In Pursuit of Creativity in Science and Technology, Outline of White Paper on Science and Technology 1982," *Science & Technology in Japan,* vol. 2, no. 6, April/June 1983, pp. 18–23.

[81] *Research and Development in Industry, 1979,* NSF 81-324, National Science Foundation, Washington, D.C., 1981, p. 11; *Academic Science: R&D Funds, Fiscal Year 1980,* NSF 82-300, 1982, p. 8.

[82] Shigeru, N., "The Role Played by Universities in Scientific and Technological Development in Japan," *Journal of World History,* vol. 9. no. 2, 1965, p. 354.

[83] "Science and Technology R&D Expenditures," *Science & Technology in Japan,* vol. 2, no. 6, April/June 1983, pp. 40–41.

[84] Bloom, J. L., "Japanese Science and Technology, The View from the Other Side," *Speaking of Japan,* Aug. 1981, p. 28.

[85] *Ibid.*

[86] *Indicators of Science and Technology—1980,* NSB-81-1, National Science Foundation, Wash., D.C., pp. 68–69, 32–33; "Summary of Data," in annual Engineering College Research and Graduate Study issue of *Engineering Education,* vol. 73, no. 6, Mar. 1983, p. 432.

[87] "The Vision of MITI Policies in the 1980s, Summary," *op. cit.*, ref. 2, pp. 15–16.

[88] For a discussion of Japanese, U.S. and European efforts to develop fifth-generation computers, see *IEEE Spectrum,* vol. 20, no. 11, Nov. 1983.

[89] Amaya, T., "Human Resource Development in Industry," The Japan Institute of Labour, Tokyo, 1983, pp. 5, 7.

[90] Kikuchi, *op. cit.*, ref. 3, p. 21.

[91] *Ibid.*, p. 12.

[92] Shimada, *op. cit.*, ref. 3, p. 21.

[93] *Education in Industry,* The Conference Board, Inc., NY, 1977.

[94] Long-Term Outlook Committee, *op. cit.*, ref. 1, p. 27.

[95] Amaya, *op. cit.*, ref. 35, p. 21.

[96] Long-Term Outlook Committee, *op. cit.*, ref. 1, p. 125.

[97] *Education in Industry, op. cit.*, ref. 39, pp. 116–117.

[98] Boyer, E., ''Start-Up Ventures Blossom in Japan,'' *Fortune*, Sept. 5, 1983, p. 118.

[99] ''All the Hazards and Threats of Success,'' special issue on Japan, *Time*, Aug. 1, 1983, p. 20.

[100] ''Japan-U.S. Trade: New Open Market Measures Help to Improve Trade Climate,'' *Fortune*, Aug. 23, 1982, p. 48.

[101] ''Japanese and the Eighties: A Period of Challenge and Opportunity,'' *Fortune*, Aug. 11, 1980, p. 29.

[102] ''Hisashi Shinto: Discovering America's Secret,'' *The Washington Post*, April 10, 1983, p. F1.

[103] The National Science Board Commission on Precollege Education in Mathematics, Science and Technology, ''Educating Americans for the 21st Century: A Plan of Action for Improving Mathematics, Science and Technology Education for all American Elementary and Secondary Students So That Their Achievement Is the Best in the World by 1995,'' CPCE-NSF-03, National Science Foundation, Washington, D.C. Sept. 12, 1983, p. v.

EPILOGUE: THE IMPACT OF CULTURAL VALUES ON ENGINEERING: A SUMMATION

Bruno O. Weinschel

In these pages we learn from leaders of high technology industries about some of the differences in business and engineering cultures among Western Europe, Japan, and the United States. They include differences in education, continuing education, attitudes, and work patterns.

Yoshi Tsurumi of the City University of New York warns us that unless corporations become adaptive to changing economic conditions and changing technologies, they may not survive. Many of our U.S., Western European, and Japanese corporations are practicing the principles that will assure survival: adaptiveness, an appreciation of the individual, participative management, decision-making at the lowest possible level, and good two-way communication.

One of the crucial differences, in my opinion, is in education. Larry Grayson, the advisor for mathematics, science, and technology at the U.S. Department of Education, has assayed comparative achievements in precollege education, especially in mathematics and the sciences. He reported that, for the last 15 years, the Japanese have always come out ahead.

The general status of Japanese education is high indeed. The time spent on mathematics and sciences through high school may be up to three times that of the equivalent subjects in the United States. This is partly a result of the longer Japanese school day, half days on Saturday, and shorter vacations, which yield approximately two more years of precollege schooling. Because of this training, the average Japanese high school graduate can understand and use the tools of statistical quality control in the factory. The United States, on the other hand, requires college educated workers for that kind of performance.

West Germany, for its part, has a highly successful apprenticeship system. About half its high school population is enrolled for a four-year term in apprenticeship systems in several hundred thousand organizations. The young people go to school one day a week and work four days. That system is the backbone of the country's work force skills. An equivalent system does not exist in the United States.

If the United States expects to improve its educational system, it will be necessary to interact locally with 50 state jurisdictions, each of which encompasses, typically, some 20 counties. This cannot be done centrally through the federal government. Nor can it be done through a mere infusion of money. Individual school boards must be convinced that it is in the interests of their communities to invest in the future and to increase both the skill levels and remuneration of teachers. The IEEE may be able to help; over 75 percent of the IEEE membership has endorsed the involvement of the Institute in educational matters at local levels through the Regional Activities board and the U.S. Activities Board, and in other ways, such as joint projects with other societies.

Manufacturing renaissance

The conference documented here also underscores the importance of manufacturing engineering. Trading partners of the United States in Western Europe and in Japan guide their most brilliant engineers into manufacturing, while U.S. universities stress untargeted research for their brightest students. We also learn from our Japanese colleagues that their research products are strongly market-driven. Product goals are decided at a high level of management. Then, subsidiary process technologies are developed to manufacture the products. Fortunately, a renaissance of interest in and understanding of the importance of manufacturing engineering in the United States is underway.

The United States must begin to practice manufacturing engineering in a much larger framework, one that ranges from the original research and development of the product and its manufacturing processes through product design, manufacturing engineering, quality assurance, and customer training and servicing of the end product. This enlarged scope in turn requires a change of emphasis and values in academia. The United States has 280 engineering schools that are approved by the Accreditation Board for Engineering and Technology (ABET). We have a faculty of about 20,000 engineering professors. Yet there are few schools that make it their business to teach the design of products that are manufacturable, reliable, high in quality, maintainable, pleasing to the user, and cost-competitive. These values must be imparted to our young graduates if the United States is to become more competitive in world trade.

One of the goals of the IEEE is to better understand the needs of continuing education. We have learned from several studies and we have heard from our overseas colleagues that the best teaching does not come from academia, but is done in industry by industry specialists. Indeed, some of our commercial developments today are well ahead of the research in academia. Some commercial developments are even ahead of defense technology, which leads to difficult problems of dual use and export control. There must be a two-way street between academia and our advanced sectors of industry. We must bring professors on sabbatical into industry while advanced industry people volunteer as adjunct professors at universities.

Robert Noyce of Intel identifies a serious problem. Dr. Noyce observes that the federal government deficit sets a bad example for the rest of the country. The deficit must be brought under control. But there is little we in the IEEE, as a group, can do about it, although we can, as individuals, exercise our influence at the polls. On the other hand, U.S. trade associations are in a position to influence specific legislation. As Dr. Noyce has properly said, our federal government does not comprehend the role of capital formation. For example, the recent reshaping of the internal revenue code has an adverse impact on industry by discouraging investment in new equipment and technologies. Trade associations such as the American Electronics Association seem the best mechanism for bringing our views to our legislators.

Quality of management

Professor Tsurumi notes that at the base of many of the problems in the United States is the quality of the industrial management. Of course we have organizations like IBM, AT&T, General Electric, and Hewlett-Packard that are managed beautifully, but there are many more that do not yet recognize that autocratic management cannot enhance productivity. We must somehow impart to those schools of business administration teaching short-term optimization and authoritarian management, that managers must know and understand what happens on the manufacturing floor. Furthermore, top managers should be literate in technology, and should recognize and benefit from the value of the individual's contribution.

The report of the President's commission studying competitiveness, chaired by John Young, the board chairman of Hewlett-Packard, made these recommendations, among many others:

• Creating a federal Department of Science and Technology
• Establishing a focal point for foreign trade by creating a federal Department of Foreign Trade and Investment.

The IEEE favors both steps, and has testified for many years before Congress in support of the latter, now embodied in a bill sponsored by Senator Roth.

We also learn from our colleagues that engineers must be closely coupled to the needs of the marketplace. We hear from Michiyuki Uenohara of NEC that approximately 80 percent of NEC's manufacturing engineers are in direct contact with the customer. We learn from Dean Morton at Hewlett-Packard that large numbers of H-P's marketing force are graduate engineers. This is probably one of the reasons why Hewlett-Packard continues to be so successful. But it is not customary in U.S. industry. We learn from our friends at Siemens and Philips that their engineers, too, are closely coupled to the marketplace.

We learn too that in Western Europe and Japan engineers change jobs less frequently, if at all. This lower mobility encourages employers to make long-term investments in continuing education. It is unlikely that U.S. engineers will change their job mobility pattern, but U.S. companies can be encouraged to train their engineers nevertheless. Toward this end, the IEEE may support a tax proposal that would make continuing education more attractive to U.S. companies by defining its cost as a tax credit.

Learning from one another

Overall, the technologically advanced nations of the world may well be learning from one another that the successful technologies used to bring products to the marketplace may not be as country-specific as we once believed. While some technologies may be limited in their transferability because of cultural differences, others may work just as well in any advanced nation, provided that informed legislators, industrialists, and others of influence help pave the way.

INDEX

EDITOR'S BIOGRAPHY

Donald Christiansen is the Editor and Publisher of *IEEE Spectrum* and chairman of its editorial board. Christiansen's early interest in science and technology was crystallized during World War II, when he served as a Navy radio technician. After the war, he earned his BEE from Cornell University.

His industrial experience was with Philco Corp. and CBS Electronics Division in research, development, design, manufacturing, and marketing. It ranged over radio, television, radar, and electronic components. A founder of the Merrimack Valley subsection of the Institute of Radio Engineers, Christiansen is a Fellow of the IEEE and of the World Academy of Art and Science. He is a member of numerous technical societies, including the New York Academy of Sciences and The Royal Institution. He is an Eminent Member of Eta Kappa Nu.

Christiansen's first job in publishing was as solid-state editor of *Electronic Design*. He later became editor-in-chief of *Electronics*, and was named editor of *IEEE Spectrum* in 1971. He is co-editor with Donald Fink of McGraw-Hill's *Electronics Engineers' Handbook*.